WILLIAM CAXTON

A QUINCENTENARY BIOGRAPHY
OF ENGLAND'S FIRST PRINTER

I Caxton presents his Recuyell of the Histories of Troy to Margaret Duchess of Burgundy.

WILLIAM CAXTON

A QUINCENTENARY BIOGRAPHY
OF ENGLAND'S FIRST PRINTER

By

George D. Painter

1976

CHATTO & WINDUS

LONDON

Published by
Chatto & Windus Ltd.
40-42 William IV Street,
London WC2N 4DF

*

Clarke, Irwin & Co. Ltd.
Toronto

ISBN 0 7011 2198 X
© George D. Painter 1976

Printed in Great Britain by
R. & R. Clark Ltd., Edinburgh

I dedicate this book, as always, to my wife Joan with love, and with gratitude for her encouragement, help, and kindness. I offer it also to the memory of the late Victor Scholderer, Allan H. Stevenson, and Christopher A. Webb, and to Leslie A. Sheppard, Howard M. Nixon, and Dennis E. Rhodes, in thanks for their friendship and example during many years as colleagues in the study of fifteenth-century printing.

ACKNOWLEDGMENTS

The author and publishers are indebted to the
following for permission to reproduce illustrations from books in their care: The British
Library for nos. III-V, 1-8. The Curators of
the Bodleian Library, for nos. IIb, VI. The
Huntington Library, San Marino, California,
for no. I. His Grace the Archbishop of Canterbury and the Trustees of Lambeth
Palace Library for no. IIa.

FOREWORD

Printing was first brought to England five hundred years ago by William Caxton, a retired merchant and diplomat in his middle fifties, who set up his press within the precincts of Westminster Abbey in the autumn of 1476. Few events in English history have been more tremendous in their significance or more lasting in their consequences for our language and literature, our daily life and culture. The quincentenary of Caxton's advent thoroughly deserves its celebration by his countrymen, his fellow Europeans, and the English-speaking world. The time seems ripe also for an attempt to assess and correct our existing knowledge of Caxton's life and work, and wherever possible to discover more.

Caxton is one of the most famous yet least known of great Englishmen. The sources for his eventful life are unexpectedly rich and varied, but have not as yet been adequately studied and interpreted. They range from the abundant contemporary documents in which his name and affairs are mentioned, to the hundred surviving editions of books from his press, with their texts, types, and woodcut illustrations, and the revealing and entertaining prologues and epilogues which he wrote for them. An attractive and original but distinctively English character emerges, a man of gusto, piety and humour, of business acumen and enduring political loyalties, a gifted compulsive writer with a special delight in printing his own works, an editor-publisher of Chaucer, Gower, Lydgate and Malory who enriched English literature by translating the bestsellers of France and Burgundy, and the offstage companion, as propagandist and protégé of Yorkist and Tudor royalty and nobility, of the heroes and villains of Shakespeare's *Henry VI* and *Richard III*.

In this biography I have described and discussed every known Caxton document and edition, both intrinsically and in relation to the events, persons, and movements of contemporary history in which Caxton was so intimately involved. I have tried to rectify the disconcertingly many established and hitherto unsuspected errors of fact or inference in the work of William Blades, E. G. Duff, W.J. B. Crotch and others, to bring new light and truth to all aspects of Caxton's career from an independent study of the primary sources, and to write for the general reader, the student, and the specialist scholar alike. New conclusions are reached on Caxton's family connections, his early activities as apprentice in London and cloth-trader at Bruges, his appointment and fall as Governor of the English merchants in the Low Countries, his diplomatic missions in the protracted trade negotiations of the 1460s, his discovery of his vocation for writing and printing, his relationships with his instructor Johann Veldener and Colard Mansion his associate, and the foundation and chronology of his first press at Bruges. I show that it was from Mansion and the Bruges scribal tradition that Caxton borrowed and adapted his practices, otherwise unique among fifteenth-century printers, of writing his own translations for publication, obtaining commissions for these and other works from royal or noble patrons, and introducing them with original prologues and epilogues as a vehicle for political or personal propaganda on behalf of his clients. Caxton's hitherto unrealised function as a Yorkist and Tudor propagandist is explored in detail as a major key to his entire career as a printer. New information is given on the sources and authorship of Caxton texts previously misattributed, and dates are supplied on new typographical and other evidence for many of Caxton's undated editions.

Fifteenth-century English spelling and punctuation have here been modernised, but without change of wording. Examples of Caxton's own practice (or his compositors') will be found in the plates and text illustrations, and in the Notes on these. All Caxton's types are illustrated in natural size. The references at the end of the volume lead primarily to printed sources used in the text, but also to other discussions of Caxton matters, even when (as is usually the case) the author cited gives a view more or less at variance with the one here propounded. I have indicated in context my grateful indebtedness to the discoveries and ideas of earlier writers such as Blades, Duff, Plomer, Crotch, and to A. H. Stevenson, L. A. Sheppard, C. F. Bühler, L. and ₐW. Hellinga, H. M. Nixon, Professor N. F. Blake, and other present-day Caxton specialists. Mr Nicolas Barker was so kind as to read my typescript at a rather primitive stage, and I am deeply obliged to him for many valuable corrections and suggestions.

I am grateful to the British Library for the professional lifetime in which work on Caxton, and association with the world's finest collection of his own and other fifteenth-century books, were among my everyday duties, and to Mr. Howard M. Nixon, Librarian of Westminster Abbey and member of the Caxton Commemoration Committee, for inviting me to write now a book which I had always intended to write some day. This biography is, of course, in no way official, and I alone am responsible for it; still, I hope it will serve the aims of the Caxton Commemoration of 1976, and contribute to knowledge and understanding of the great man who is being commemorated.

HOVE 1974-6   GEORGE D. PAINTER

CONTENTS

1 The Family   1

2 The Apprentice   8

3 The Merchant   16

4 The Governor   25

5 The Diplomat   32

6 The Exile   43

7 Learning to Print at Cologne   51

8 Printing at Bruges 1: Margaret and Clarence   59

9 Printing at Bruges 2: Colard Mansion   72

10 A Shop in the Abbey   82

11 At the Red Pale   98

12 Friends in Court   108

13 Under which King?   121

14 Orders of Chivalry   141

15 With a Strange Device   151

16 The Two Queen Mothers   163

17 The Art of Dying   173

Continued overleaf

CONTENTS

Notes on the Illustrations   193

Select Bibliography   196

References   203

Chronological List of Caxton's Editions   211

Caxton's Device as Evidence for Dating   215

Index   217

LIST OF PLATES

(See also *Notes on the Illustrations*, p. 193)

I Caxton presents his *Recuyell of the Histories of Troy* to Margaret Duchess of Burgundy   *Frontispiece*

IIa Earl Rivers presents his *Dicts of the Philosophers* to Edward IV   84

IIb Caxton's Type 3, from his *Advertisement for the Sarum Ordinal*, 1479   84

III Caxton's Type 1, from *Recuyell of the Histories of Troy*, 1475   85

IV Caxton's Type 2, from *Dicts of the Philosophers*, 1477   100

V Caxton's Type 7, from the 1489 *Indulgence*   101

VI Caxton's Type 8, from *Ars moriendi*, 1491   180

ILLUSTRATIONS IN TEXT

1 Death among the Printers. The earliest known illustration of a printing press, from *Danse Macabre*, Lyons 1499.   55

2 Caxton's Type 2*, from *Mirror of the World*, 1481.   109

3 Caxton's Type 4, from *Knight of the Tower*, 1484.   127

4 Caxton's Type 4*, from *Canterbury Tales II*, 1483.   133

5 Caxton's Device, from *Legenda ad Usum Sarum*, 1488.   159

6 Caxton's Type 5, from *Doctrinal of Sapience*, 1489.   171

7 Caxton's Type 6, from *Eneydos*, 1490.   175

8 Caxton's last book, *Fifteen Oes*, 1491.   185

1

THE FAMILY

EVERY biography ought to begin, of course, with an account of the great man's parents and pedigree, and the date and place of his birth. Alas, it is the first duty of a biographer of England's first printer to confess that he can give none of this information! No records yet discovered have identified William Caxton's father or mother or family. The time of his birth can only be conjectured with a possible error of ten years. We have his own word for the county, but not the place, where he was born and bred: 'I was born and learned mine English in Kent in the Weald,' he says in the prologue to his first book, *The Recuyell of the Histories of Troy*. His parents saw to his education: 'I am also bounden to pray for my father's and mother's souls,' he remarks in his prologue to *Charles the Great*, 'that in my youth set me to school, by which by the sufferance of God I get my living I hope truly.' And this is all, absolutely all the certain knowledge we possess about William Caxton until his middle teens, when archival sources begin to fill the gap.

Still, we need not give up all hope of discovering something about the Printer's family. Documents from his lifetime and many years before reveal the existence of an astonishing multitude of persons who shared his surname. In my sceptical youth I felt obliged to ignore every member of these apparently chaotic coveys of Caxtons, for want of positive evidence that any individual or group among them was genuinely connected with the Printer. But now, in my credulous old age, it seems worth while to try again; I find it increasingly likely that some or most of these real though shadowy namesakes belonged to one and the same expanding, rising and falling family; it appears encouragingly possible to find traces of significant order in their geographical and chronological distribution, and even here and there to detect symptoms of actual kinship and personal intercommunication with the Printer. So I will now attempt to guide myself and the reader through this baffling but fascinating space-time labyrinth of Caxtons, which ramifies throughout the southeast quarter of England, including Norfolk, Suffolk, Essex, London, and Kent, and through more than two hundred years from the end of the thirteenth century to the beginning of the sixteenth.

Luckily Caxton's name itself provides a useful clue. It would have been unhelpful if his first ancestor to bear a hereditary surname (as became increasingly usual from the twelfth century onwards) had been called Smith, Miller or Painter from his trade, or Johnson, Thomson, Wilson from his father's christian name, or White, Brown, Little from his complexion or physique, or Ford, Hill, Heath from the neighbourhood of his house. Such names were common to thousands of different and unrelated families all over medieval England. Caxton's surname, however, is more helpful; it is a place-name, and means that his forebears came from a place called Caxton or the like. But which?

There is a Caxton in Cambridgeshire, near a crossway on the Roman road of Ermine Street nine miles west of Cambridge, where to this day one of the last gibbets in England looms squat and weather-blackened, once a good pull-up for highwaymen. The derivation seems uncertain,[1] but the name and spelling fit our man, and have led some enquirers to favour this Cambridgeshire village as the starting-point of the Printer's stock. But is the spelling really relevant? Medieval Englishmen wrote their own language phonetically, and were content to spell many words, especially proper names, in any way that reproduced the sound they spoke or heard. The same person is often spelt Caxton, Caston, Cawston, Causton and so on, sometimes in one and the same document.[2] The 'a' was pronounced broad, as in 'talk' or 'walk', rather than short as in 'cat', and 'xt' and 'st' were virtually indistinguishable and interchangeable. Even the Cambridgeshire village of Caxton is already spelt *Caustone* when it first appears, in Domesday Book in 1086. Hence the spelling fails to provide an argument for Caxton (Camb.), and as far as the records show none of our Caxtons or Caustons seems to have come from there. So I am inclined to propose a different candidate, the village of Cawston in Norfolk, eighteen miles northwest of Norwich.

Cawston (spelt *Caustuna* in Domesday Book, *Causton* in 1159) was a busy weaving town in the later middle ages, famous for its own brand of woollen cloth called 'cawston'. On general geographical and occupational grounds Cawston (Norfolk) seems a more likely place of origin for the Caxtons or Caustons who spread southward to Suffolk and Essex, settled in the clothmaking county of Kent, and founded a whole dynasty of cloth-merchants or mercers in London. But one piece of positive evidence, which no one seems to have taken into account, directly connects the Caustons of London, and consequently (as we shall see) the Caxtons of Kent and Westminster,

---

[1] Working from the later, twelfth-century forms *Kachestone*, *Cakeston*, the great etymologist Skeat suggested an original *Cahestun*, meaning the farm of an Anglosaxon settler named *Cah*; but the equally great Ekwall preferred a derivation from the Scandinavian personal name *Kakkr* (with which the chief present-day specialist P. H. Reaney concurs), or else from *kex* meaning umbelliferous plants. There is also a Cawston in Warwickshire near Rugby, and another in Shropshire near Clunton and Clunbury, the quietest places under the sun, but neither has any apparent connection with our Caxtons.

[2] 'The actual differences in the spelling of the proper name are of no account,' remarks Crotch (p. xxx), 'for William Caxton himself appears as Catston, Caxston, Caxtun, and Kaxsum; Thomas Caxton as Cawston, Causton, Caueston, and John de Cawston as John Cawystin or Caxton.' It will be seen that our further enquiries add a Robert 'Caxton, alias Cawston', an Oliver 'Caston, alias Causton, alias Caxton', a Richard 'Caxston or Caston', and another William 'Caxton otherwise called Causton' to these examples.

with Cawston in Norfolk. This missing link is an entry in the Hustings Rolls of the City of London early in the fifteenth century, referring to 'Isabella, wife of Thomas Huchons of Causton in the county of Norfolk, cousin (*consanguinea*) and heiress of William Causton citizen and mercer of London'. Evidently Isabella belonged to a branch of the family that had remained at Cawston, Norfolk, or at least kept sufficient ties with the old home to marry a husband from there. However this may be, the connection of the London Caustons with the Norfolk town is now established.

This William Causton died, and cousin Isabella inherited, a little before 1406, after a long life, for he was apparently a son of the wealthy William de Causton who died in 1354. Young William had taken the livery of the Mercers' Company in 1348 along with a whole bevy of other de Caustons, Richard, Michael, Henry, Theobald, Nicholas, and Roger. Old Father William was himself a London mercer and perhaps the richest of all the family, with lands at Edmonton, Enfield and Tottenham, as well as many tenements in various parts of the City. Nearly 350 documents survive concerning his properties and transactions, with mentions not only of young Henry and Nicholas de Causton, but also of old William's contemporary (and brother?) John de Causton, mercer and sheriff of London, who had made himself famous in 1326 when he boldly refused to give the King's secretary lodging in his house in Billingsgate, and got away with it.[1] In his will dated 1353 John established a chantry in the convent of Holywell outside Bishopsgate for masses to be sung for his own soul and old William's. A still older generation of Caxtons and Caustons (the spelling, as usual, seems to be immaterial) had preceded father William in London, for wills exist, each bequeathing substantial house-properties in the City, of a William de Causton in 1297, Alexander de Causton in 1299, William de Caxton in 1311, and Matilda de Caxton in 1342. The London dynasty of Caxton-Caustons continued well into the Printer's lifetime, and kept their connection with the Mercer's Company. Yet another William Causton was apprenticed to Thomas Gedeney, mercer, in 1401, and a Richard Caxton completed his apprenticeship to John Harowe, mercer, in 1447. A William Causton tailor is mentioned in 1435, 1448, and 1462, and a William de Caxtone was assessed for the minute levy of one penny on a tenement in 1444. An Oliver Causton was appointed collector in the ward of Cripplegate Without on 12 February 1481 for Edward IV's 'benevolence' in the war against Scotland. Perhaps the family became a little less prosperous, for one

---

[1] Edward II arrived in July with his court to spend summer in the Tower of London. John de Causton wiped out the chalk-mark for the secretary's billeting on his door, and drove his retinue and their horses away, whereupon the billeting sergeant sued him for £1000 damages. The Mayor of London appeared in person for the defence, producing a charter of the King's grandfather Henry III forbidding compulsory billeting within the City walls, together with Edward II's own confirmation of this and other privileges of the City. Verdict: Not Guilty.

or two seem to have fallen on evil days. Roger in old age received alms of thirteen shillings from the Mercers' Company in 1395, and a Stephen Causton, who took the mercers' livery in 1424, was granted alms for forty-five weeks at fourteen pence a week in 1434.

Meanwhile Caustons and Caxtons related to the London family had settled in Kent, perhaps in consequence of Edward III's successful efforts to encourage the clothmaking industry in that county. The wise monarch, noticing (as Fuller of Fuller's *Worthies* remarks) that 'Englishmen knew no more what to do with their wool than the sheep that wore it'—invited eighty families of Flemish weavers to settle in Kent in 1331, and in 1338 prohibited export of English wool and wearing of foreign cloth; whereafter for 400 years, until the dawn of the Industrial Revolution, Kent became one of the chief centres for the weaving, fulling, and dyeing of wool. Hugh de Causton of the City of London held the manor of Caustons in the Kentish Weald near Hadlow in 1370. He alienated it not long after to the Wattons of nearby Addington, so the old idea that the Printer was born at Hadlow is groundless; but Hugh's kindred Caustons and Caxtons spread all over the county during the next hundred years and more, with centres at Canterbury, Sandwich, Romney, Bircholt near Ashford, Tenterden and elsewhere.[1] A Walter Causton was monk and precentor at Canterbury Cathedral in 1383, and prior of St. Martin's, Dover, in 1392. At least three of the Kentish Caxtons kept up the London branch's affiliation with the Mercers' Company; for a William Caxton who took the freedom of Canterbury in 1431, Hugh Caxton active at Sandwich in the 1450s, and John Caxton who died 12 October 1485 and was buried at St. Alphege's in Canterbury (perhaps a son of 'Robert Caxton, alias Cawston', a property-owner in the same parish who died in 1459), were all mercers. There were also at least two other Kentish Williams besides the one just named, one a native of Romney who gave land to a Sandwich charity in 1405, while the other, of Westerham, left a will dated 1485. Probably yet another (unless he is the Canterbury William or the Romney William turning up in a different place) is the William Caxton named for a largish sum (nine shillings) in the subsidy rolls for the hundred of Shamwell, apparently at some time in the 1420s. Shamwell hundred was a narrow strip reaching from the Medway at Strood opposite Rochester to the Thames east of Gravesend. Perhaps the memory of Shamwell cousins underlies Caxton's little joke in *Golden Legend* when he denies the slander that Kentishmen have tails: 'it is said commonly that this befell at Strood in Kent, but blessed be God at this day is no such deformity'.[2]

Perhaps the most intriguing of all is Thomas Caxton of Tenterden,

[1] Halfway between Cranbrook, where the weaving industry founded by Edward III lingered on until the early nineteenth century, and Tenterden is a tract of ancient woodland called Causton Wood to this day, and so marked on Ordnance Survey maps.

[2] For Caxton's sensitivity about tails in Kent see also p. 26, footnote 1.

or rather the two Thomas Caxtons of that Wealden town. The Thomas Caxton who paid substantial property taxes at Bircholt near Ashford in 1417 must be the same as the one mentioned in Tenterden documents of 1418, 1428, and 1439; but he can hardly have been born later than the 1390s, so he cannot possibly be (as Crotch thought) the same as the Thomas Caxton who first appears in records at Lydd in 1454 as 'of the parish of Tenterden', and died at Lydd in 1495. This second Thomas Caxton, a prominent lawyer and business man specialising in the affairs of the Cinque Ports, was successively or concurrently Common Clerk to Lydd, Romney, and Sandwich from 1468 to the 1480s. We find him obtaining a letter of general pardon, just as the Printer did, after the Kentish rebellion of 1471, sitting on a commission of enquiry into wool shipments to the Staple of Calais in 1474, and riding to Sandwich on an undated occasion for an interview with the King's brother-in-law and the Printer's patron, Anthony, Earl Rivers.[1]

The family connection of the Thomas Caxtons and other Kentish Caxtons with those of the London dynasty is certain from the remarkable but not isolated fact,[2] that the first Thomas Caxton of Tenterden is mentioned, as witness to a will dated 28 May 1439, in the vast archive of 346 documents relating to the properties of old Father William de Causton the Londoner (who died in 1354) and his son William (who died before 1406). It is equally significant that these documents have been kept since time immemorial among the muniments of Westminster Abbey, although they have nothing to do with the properties or business transactions of the Abbey itself. They were discovered in 1893 by E. J. L. Scott, who suggested that they were deposited there by the Printer himself. But when we find a whole subgroup of the London Caxtons associated with Westminster during a period from well before the Printer's arrival there in 1476 until well after his death in 1491, it seems more likely that the depositor was one of these—preferably Richard Caxton (also called 'Caxston or Caston'), one of the last of the family, who was a monk of the Abbey from 1473 until his death as Sacrist in 1504. Oliver Caston, 'alias Causton, alias Caxton', citizen and Skinner of London, was buried at the Westminster parish church of St. Margaret's in

---

[1] An alleged Thomas 'Cacston or Causton' twice cited from the Mercers' records by Blades is unfortunately a mere misreading for Thomas Sawston, who has nothing to do with the Caxton family. This Sawston appears in a list of mercers chosen by the Mercers' court of 5 June 1461 to ride to meet Edward IV on his pre-coronation entry into London (26 June), and again on 10 March 1463 to meet the King 'coming from the north' (Blades garbles these entries); he was 'common linen measurer' in 1478, and was replaced 'in great disease' in 1480.

[2] As noted by Crotch (pp. xxxiii, cvii). The connection between the London and Kentish Caxtons within the Printer's lifetime is further shown by a hitherto unnoticed entry in the Plea and Memoranda Rolls of 10 March 1453, in which Richard Caxton the London mercer (who had completed his apprenticeship to the Printer's associate John Harowe in 1447) is found in a transaction with Hugh Caxton the mercer of Sandwich.

1465; so was a William Caxton in 1478; a John Caxton became a member of the same church's Guild of Our Lady (a religious, charitable and social fraternity for lay parishioners) in 1474–7; and we shall find the Printer a prominent member of the same guild and buried in the same church.

Our tally of Caxtons is still not quite complete. The will exists, dated 1490, of a John Caxton of Hadlow Hall, Essex; evidently this John had remembered with pride the name of the family manor of Caustons near Hadlow in Kent, over a century after it had passed to strangers. Yet another lineage of Caxtons were merchants during the fourteenth and fifteenth centuries at Norwich, where they used the punning trademark of three hot cross buns and a winebarrel (*cakes-tun*). A Roger de Cawston was Town Clerk there in 1322, and a John Cawston left bequests to religious foundations at Norwich in 1466. Lastly we have the strange case of the Caxton Deeds, which were discovered in 1923, acquired by the press magnate Lord Kemsley in the belief that the Printer himself was mentioned in them, and generously presented in 1962 by his successor Lord Thomson to the Department of Manuscripts in the British Museum (as it was called at that time). These fifteen deeds relate to the manor of Little Wratting in Suffolk during the period 1420-67, under its owners Philip Caxton and his wife Denise and their sons Philip and William. In fact this William cannot possibly be the same as the Printer, who was a Kentishman, for the Little Wratting man was already of age in 1438, and is described in 1457 (when the Printer had long been a Mercer resident at Bruges) as 'William Caxton otherwise called Causton, Saddler'. Even so, a surprisingly large number of witnesses and other parties to these deeds were connected more or less distantly with the Printer;[1] so once again we have the impression of different branches of the same family communicating through a common circle of acquaintances.

Our survey has revealed the existence of forty individual male Caxtons, not counting the Printer, of whom at least sixteen belonged to the Mercers' Company, and twelve (including five mercers) were named William. Their distribution in time, place, and occupation evidently forms a meaningful pattern. Indeed, there now seems no reason to doubt that most if not all of these Caxtons, and the Printer

---

[1] These include Robert Large, the London mercer to whom the Printer was apprenticed; Richard, Duke of York, father of his patroness Margaret of Burgundy; Benedict Burgh, high canon of St. Stephen's Westminster and author of the translation of Cato's *Distichs* which Caxton thrice printed; Henry Lord Bourchier, later Earl of Essex, for whom as Caxton mentions Burgh wrote this work; and a Robert Burgh and a Richard Bumpstead who might conceivably be related to the Richard Burgh and Henry Bomsted who were business acquaintances of the Printer. A Philip Caxton was attorney in 1420 to Thomas, Duke of Exeter, in his capacity as guardian to the then under-age John de Vere, 12th Earl of Oxford; this 12th Earl is mentioned in a Little Wratting deed of 1456, and his son the 13th Earl was a patron of Caxton the Printer. Little Wratting is near Haverhill in the extreme southwest corner of Suffolk, just opposite Helion Bumpstead and Steeple Bumpstead (beautiful names!) in Essex.

himself, were members of a single widespread family, issuing from Cawston, Norfolk, towards the close of the thirteenth century, and flourishing for two hundred years as clothtraders, mercers, property-owners and professional men, with main and intercommunicating branches in London (later including Westminster) and Kent. The evidence does not positively identify the Printer's parents or birth-place. Tenterden, however, a weaving town in the heart of the Kentish Weald, with a Causton Wood near by, would fit his own statement perfectly, and it would not be surprising if the elder Thomas Caxton of Tenterden was his father, and the younger one his brother. Even his christian name William was apparently due to a family preference! We can see indications that some of the personal relationships that influenced his entire career, with men of the merchant classes, the Church, the nobility, the court, and royalty, were thanks to his family contacts. Above all, we can be sure that three crucial events of his life were not coincidental, but direct consequences of the proven interconnection between the Caxtons of Kent, London, and Westminster. He was born and bred a Kentish-man because of the Caxton presence in Kent. His parents made him a mercer because of the Caxton tradition, then long established and still thriving, of mercership in London. And one salient reason why William Caxton chose to set up the first English printing press in Westminster Abbey (although we shall find other motives also at work) must have been that he had relatives already settled there to welcome him.

2

THE APPRENTICE

THE first documentary evidence for an event in Caxton's life is the entry in the annual Wardens' Accounts of the Mercers' Company for the year ending 24 June 1438, recording the payment of the enrolment fee of two shillings for his apprenticeship to Robert Large. It is from this document, with the help of a bequest in Large's will which shows that Caxton was still his apprentice in April 1441, and a judgement in a lawsuit at Bruges on 2 January 1450 which shows that Caxton was then already a merchant there doing business for large sums of money, that we have to calculate his approximate age and year of birth. It needs to be pointed out that even this fee entry does not provide a hard and fast date, although some biographers have confidently stated that 'on the 24th June 1438 Caxton was entered as one of Robert Large's apprentices.' The Wardens' Accounts, which survive only for the period 1391-1464, were made up yearly to St. John the Baptist's Day, 24 June, and recorded merely the sums received during the previous twelve months for fees or fines, etc., without specifying the exact date of each transaction. Hence Caxton's enrolment fee may have been paid at any time from 25 June 1437 to 24 June 1438, and the same proviso applies, of course, to all other entries for the enrolment or issue of apprentices and so on, including those to be cited here. For convenience, however, we may as well refer to Caxton's year as 1438, and likewise for the rest, with a fifty-fifty chance of being right.

It was argued by Blades, followed by Duff, Plomer, Crotch, and others, that apprentices usually issued at the age of twenty-four after serving from seven to ten years, the minimum age of entry being fourteen; hence Caxton was between fourteen and seventeen years old in 1438, and must have been born within the period 1421-4. But Professor N. F. Blake has investigated the matter much more deeply, with results that throw doubt on the reliability of all these premisses. He shows that mercers, including Large himself, were often lax in the payment of enrolment and issue fees for their apprentices. Sometimes no entry was made at all: for example, there is no record of Caxton's issue, and both the enrolment and the issue of Robert Dedes, who is mentioned among Large's apprentices in his will, remain unregistered. No issue is recorded for 411 out of 670 apprentices enrolled in the period 1423-52; and although this is not quite 'an overwhelming proportion', as Blake suggests, it is certainly a substantial one, being more than half but not quite two thirds. Sometimes the fees were only paid when long overdue, so that on various occasions[1] enrolment and issue fees were paid simultaneously. So, if we relied blindly upon the Wardens' Accounts, we should reach the absurd conclusion that an apprenticeship could last for any period from a minimum of no time at all (as in the case of Large's apprentice Christopher Heton, for whom both fees were paid in 1443, two years after Large's death) to a maximum of twenty-one years (as in the case of Large's apprentice John Harowe, whose enrolment fee was paid in 1423, though his issue payment was not made until 1443-4).[2] Blake further shows that the age of twenty-four was not a fixed point for the end of service, for cases are found of an apprentice issuing as young as twenty-one or as old as twenty-six. He concludes that we can be sure of no more than that Caxton was born between the extremes of 1415 and 1424. That is, if Caxton was enrolled at the latest possible date (1438) and at the earliest possible age (14), he would have been born in 1424; while if he issued at the earliest possible date (1441) and at the latest possible age (26), he would have been born in 1415.

Blake's arguments, however, though correct in logic and incontrovertible, show only the possibility of doubt. They consist partly of a *reductio ad absurdum*, partly of exceptional cases, and hence omit to consider the realities or likelihoods of actual life. If we take another look, it seems statistically clear from Blake's own figures, which show a significant concentration of issues within a median duration of service, that the normal period of apprenticeship lay between seven and ten or eleven years, and also that delay in enrolment, at any rate beyond a year or two, was not usual.[3] Other evidence points the same way. The Mercer's oath at this very time included a vow 'to take none apprentice for less time than seven

---

[1] Blake says 'often', but his tables show this happening only on 19 occasions out of 259.

[2] In fact John Harowe must have issued long before 1443, as in 1446-7 he paid the issue fee of his own apprentice Richard Caxton, having been already master of two apprentices in 1443. Blake inadvertently but repeatedly (four times) gives Harowe's issue-date as 1454, though rightly calculating his ostensible period of service as twenty-one years. Blades gives the correct date, 1443-4, for the registration of Harowe's issue fee, but he and Crotch are misled by the timelag into dividing this one and only John Harowe into two persons, father and son.

[3] Of the total number of issues 63 per cent are found in these median ranges. But evidently the greater part of the 23 per cent registered as serving impossibly or improbably short terms from no years to six must be a figment due merely to delay in registration, and must be incorporated in the median ranges, which will thus total about 86 per cent. An ostensible service of more than eleven years is only given for 14 per cent; and some, if not most, of these cases must again be unreal and due to delay in fee-paying. It follows that (*a*) the norm of seven to eleven (or rather ten?) years is real, and (*b*) that any long delay in registration was exceptional, and is easily detectable. It may be noted that the figures are also slightly distorted by the fact that the true year dates run from 25 June in one year to 24 June in the next, so that the convenience-date causes a 50 percent bias towards a notional dating of both enrolment and issue which is a year later than the real date.

years', 'to do him be enrolled within the first year', and 'to make him free at the term's end if he have well and truly served you'. Other records show that admission under the age of fourteen was illegal. It seems clear that apprenticeship was geared to a normal issue-age of twenty-four; the minimum service of seven years was intended to accommodate latecomers, and a ten-year stretch was for those joining at the minimum age of fourteen. Twenty-four was in fact the age of civic majority, and Large himself provides in his will that his three under-age sons must not receive their inheritances until each 'shall arrive at the age of twenty-four years'. Each of Large's three apprentices for whom credible dates of both enrolment and issue are given (Streete, Bonifaunt, and Nyche) served not more than ten or eleven years, perhaps less. In his will made 11 April 1441 Large evidently lists his six then remaining apprentices in order of seniority, and the sums of money bequeathed to each are mostly graduated in the same order. Caxton's name comes last of all, after Robert Dedes and Christopher Heton, and he receives a smaller amount, 20 marks, than anyone but Dedes (the one who was never registered at all), who gets the same. So we can infer that Caxton was the newest recruit, and that his enrolment date of 1438 was either real, or not delayed by more than a year or two. Caxton's first known appearance as a thriving business man as late as 1449 likewise suggests that his service still had several years to run in 1441.

All these considerations seem to corroborate the premisses of Blades and the rest, and to suggest that Caxton's career was not among the infrequent exceptional cases, but followed the norm. So one may conclude that it is admittedly not certain, but *is* highly probable, with a likelihood of something like seven to one, that Caxton began service in or not more than a year or two before 1438, that he was not less than fourteen or more than seventeen years old at the time, and therefore that he was born between 1420 and 1424. Crotch, when he split the difference and suggested 1422 as 'the least unlikely guess', can't have been far wrong.

The Mercers' Company was one of the oldest and wealthiest of the London guilds, with origins reaching back before the beginning of the fourteenth century, when it is already found with its own livery and priest. Its members were specialists in the wholesale export of woollen cloth, but not exclusively so. They might deal also not only in other textiles, including silks and haberdashery, but in almost any profitable ware, and we shall find Caxton on occasion concerned with pewter vessels and with iron; similarly woollen cloth was sometimes exported also by Drapers and Haberdashers, and even by Grocers and Skinners. The Mercers traded chiefly with the Low Countries, especially Flanders, for French outlets were restricted by the losing Hundred Years War, the Staple of Calais was reserved for unwoven wool, and further east on the German and Baltic seaboards the jealous Hanseatic League strove to keep a monopoly. Throughout the fifteenth century their prosperity was still rising on the wave of an economic revolution, in which English wool exports were made less and less in raw wool, and increasingly in the more profitable form of manufactured cloth. By an economic paradox (it is always sensible to send coals to Newcastle or owls to Athens if you can be sure of underselling the locals) weaving was the major industry of the mercers' best market, Flanders; but the Flemish weavers were dependent on imported English wool from Calais, under ever heavier duties, whereas the English cloth trade produced its own raw materials, and so competed favourably.

The alderman Robert Large was among the most prominent of mercers and city men. He was elected one of the four annual Wardens of the Mercers' Company in 1427, became a Sheriff of London in 1430, and in 1439, not long after young Caxton joined his service, was made Lord Mayor. The delighted mercers rode in full livery in his procession to Westminster on 29 October (those who could not be bothered were fined), and paid £5 6s 8d for the hire of sixteen trumpeters on this great occasion.

Large's house stood on the corner of two streets that still exist, Lothbury and Old Jewry, the main street of the medieval ghetto from which Edward I had expelled the Jews long ago, in 1290. The house was still there, though a little gone down in the world, when Stow, in his *Survey of London* (1598), described it as 'sometime a Jews' synagogue, since a house of Friars, then a Nobleman's house, after that a Merchant's house, where Mayoralties have been kept, but now a wine tavern' (the Windmill). The whole district was destroyed, of course, in the Great Fire of 1666, and again in the bombing of 1940-1, and again by postwar development and concrete ziggurats. But many of the medieval streets remain unchanged in name and position, and one can still stand on the corner where Large lived, or enter a church he worshipped in, St. Margaret's Lothbury (as rebuilt by Wren, just behind the Bank of England), or walk down the next street westward from Old Jewry, Ironmonger Lane, where the Mercers' Hall (which was rebuilt in 1517-52 and burned down in 1666 and 1940) lay conveniently near at the junction with Cheapside, in premises which the Mercers had bought in 1407 from the Templar's church of St. Thomas of Acon.

Large's household consisted of his second wife Johanna, his four under-age sons Thomas, Robert, Richard, and John,[1] his two daughters Alice and Elizabeth, and at least seven apprentices. These, in order of seniority, were Thomas Nyche, Richard Bonifaunt, Henry Dukmanton,[2] Robert Dedes, Christopher Heton, John Large, and

---

[1] John Large is generally assumed to have been Large's son, and very likely was so, but there seems no positive evidence that he was not a mere nephew or other relative. He must have died before 1441, as he does not appear in Large's will; indeed, he is known only from the enrolment fee entry in the Wardens' Accounts for 1437-8: 'Item John Large, Item William Caxton, the apprentices of Robert Large, 4 shillings'. This does not necessarily show that they began service on the same day together, but suggests that John Large was senior to Caxton.

[2] Misread by Blades as 'Onkmanton'.

11

Caxton himself. These, excepting John Large, are all mentioned in Robert Large's will in 1441. There may perhaps have been one or two more in 1438, for Robert Halle, Randolph Streete, and James Heton, whose enrolment fee dates range from 1427-8 to 1430-1, could possibly have remained temporarily, either as apprentices awaiting freedom, or as 'servants' (as freeman employees were called).[1]

So the boy Caxton began to learn his first trade, and to acquire the business experience and personal contacts which were to profit him thirty-five years later in his second trade of printing. It was doubtless not by mere chance that his parents were able to set him to a master of such pre-eminent status and prosperity; and Large, who could afford to choose, must surely have consented to take him on because he knew the Caxtons as an old and respected mercering family, perhaps because young William already showed promise. Obligations between master and apprentice were mutual. A master was bound by the Mercers' Ordinances to 'feed, clothe and teach well and truly his art and craft'. An apprentice swore to obey his master, serve him well, keep his trade secrets, and do no business on his own account.

During his time of apprenticeship Caxton saw and remembered small events in the streets of London, and heard of great events in the world. He later printed editions of two historical works, *Chronicles of England* and Ranulph Higden's *Polycronicon*, both of which include a continuation up to the year 1461 which is often supposed, quite wrongly, to be Caxton's own composition. In fact he reproduced *Chronicles of England* from a manuscript which already contained this continuation, and reprinted much the same text in *Polycronicon*, with a few editorial additions which he took from other sources but did not write himself. Even so, when Caxton read and printed it this text must have held special meanings for him, and revived old memories of happenings he had witnessed in person or heard the news of in youth; for the continuation, though its basic contents are a year by year account of English and European history, is also laced full of vivid and entertaining London incidents of purely local interest, so that it must have been written by a Londoner, or at any rate by a compiler using a London chronicle among his sources. In 1422, perhaps the year of Caxton's birth, 'the weather cock was set upon Paul's steeple'. In 1437 'died all the lions in the Tower of London, the which had not been seen many years before out of mind',[2] and on 14 January

---

[1] William Halle (mentioned in Large's will as 'mercer, lately my servant'), John Harowe (enrolment fee paid in 1423), and Thomas Staunton (Large's brother-in-law and executor, enrolment fee as his apprentice paid 1422-3, issue fee 1434-5) would presumably have already left. Blades says 'eight apprentices', listing the six in Large's will and adding Harowe and Streete, but forgetting John Large; Crotch accordingly says 'nine'. Blades also includes two 'servants'; but it is clear from the context that the two servants mentioned in Large's will immediately after the apprentices are only domestic servants.

[2] The menagerie in the Tower was started off by Henry III in 1235 with three leopards, a polar bear and an elephant, and remained one of the favourite sights of London until the collection was moved to the new Regent's Park Zoo in 1835—hence the proverbial expression, 'seeing the lions'.

1438 'fell down the gate with a tower on it on London Bridge toward Southwark with two arches and all that stood thereon'. Robert Large left 100 marks to the Bridge repair fund in his will. In 1441 'the Witch of Eye beside Westminster' was not so lucky as the heroine to whom Christopher Fry gave her beautiful surname, for this lady *was* for burning: 'and Margery Jourdemayne was brent in Smithfield,' says the chronicle.

The witch trials of 1440-1 were politically directed, for medieval witch hunts were exactly like our own, except that the victims were accused of real witchcraft. When Eleanor Cobham Duchess of Gloucester was condemned in November 1441 to walk barefoot down Cheapside in penance, past the mercers' shops, with the Mayor and guilds as official witnesses, everyone knew that her disgrace was intended to strike her husband Good Duke Humphrey, the King's uncle and leader of the war party, and that the time had come to run down the Hundred Years' War. The chronicler was inclined to pinpoint the first downward turn of the war to 1428, when 'the good Earl of Salisbury Sir Thomas Montagu laid siege to Orleans, at which siege he was slain by a gun that came out of the town . . . sith that he was slain Englishmen never gained nor prevailed in France but ever after began to lose little by little'.

Robert Large made his will on 11 April 1441, no doubt knowing that his end was near, for he died on the 24th, and was buried beside his first wife Elizabeth, *née* Staunton, the mother of all his children, at St. Olave's church in Cheapside.[1] Large, if any man can be judged from his last will and testament, was a kind and decent citizen, comfortably rich without being disgustingly so. He gave away nearly a quarter of his fortune to his friends, apprentices, and domestic servants, and to a sensible choice of charities (though these, of course, were intended to have a beneficial effect on his soul in the next world), including his local churches of St. Olave Cheapside and St. Margaret Lothbury, the new London conduit, the cleansing of Walbrook, poor maids and manservants desirous of marriage, bedding for St. Bartholomew's Hospital and various leper-houses, and even comforts for 'the poor prisoners in Newgate'.[2] The rest, a little more than £6000 from a total of not quite £8000, went to his family, mostly to his three sons (£1000 each) and his second wife Johanna (her dowry of 4000 marks).[3] Thomas Nyche, who had

---

[1] St. Olave's was destroyed in the Great Fire, rebuilt by Wren, and demolished in 1888.

[2] Large's bequests to the parish churches of 'Shakeston', 'Aldestre' and 'Overton', where his father and other ancestors were buried, show that he came from the Warwickshire-Leicestershire border, for these places are the modern Shackerstone, Austrey, and Orton-on-the-Hill, all within a few miles east of Tamworth. He left £100 to poor servants in Leicestershire and Warwickshire. Blades misreads Leicestershire (*Leicestr'*) as Lancashire (*Lancestr'*).

[3] The mark was two thirds of a pound, or 13s 4d. Any attempt to express medieval gold-based money in terms of our inflated modern paper is almost meaningless, but certainly one would have to multiply by a factor of more than a hundred.

ended his apprenticeship in 1438 but stayed on as an employee, was
left 50 marks; so was Bonifaunt; Dukmanton got £50, Dedes 20
marks, Heton £20, and Caxton, junior of all, 20 marks. Unfortunately the penultimate leaf of Large's will, in which he would no
doubt have provided as legally bound for the future employment of
his apprentices, is lost.[1] Possibly his executors may have transferred
Caxton to another master; but more probably Caxton remained while
Large's business was carried on by his widow, pending the coming of
age of his eldest son Thomas.

Not all went well with the Large family. Thomas and Robert
died under age, leaving Richard to inherit their shares in 1444, when
he reached the age of twenty-four; yet no money changed hands until
four years later, when Richard signed a quittance on 5 December 1448
declaring himself 'content and gratified' to receive only £1000 from
his dead brothers' portions. What happened to the other £1000?[2]
Alice died too. Elizabeth married Thomas Eyre, a Lord Mayor's son,
who received her dowry in 1446. The bereaved Johanna Large,
perhaps from grief though more probably as a stratagem for securing
her widow's rights, took the veil and ring in a vow of lifelong
chastity, but changed her mind in 1444, when Richard came of age, by
marrying John Gedney, Draper and Mayor in 1427.[3] The Church
was not pleased. 'This Godnay [sic],' says Stow, 'married the widow
of Robert Large late Mayor, which widow had taken the mantle
and ring, and the vow to live chaste to God term of her life, for the
breach whereof, the marriage done, they were troubled by the
Church, and put to penance, both he and she.'

It is not unlikely that the change-over of 1444 was the occasion of
young Caxton's sojourn at Bruges, the centre of the mercers' cloth
trade with Flanders. If so, it would not necessarily follow that his
period of service had ended. The custom was for the promising and
ambitious apprentice to go overseas in his final years, both to complete the last stage of his training, and to act on behalf of his London-
based master, like the Childe of Bristol in the ballad.[4] The Mercers'

[1] Blades printed the close of Large's will—concerning the rights of Richard
Turnat (his wife Johanna's son by a previous marriage) in his manor of Horham
in Essex, near Thaxted—only in his 1877 and 1882 editions. Unfortunately in
both these editions Blades gives the text of Large's will higgledy-piggledy,
reprinting the page contents of his 1861 edition in the order 1, 5, 4, 3, 2, 6!

[2] Large's will allowed his widow, as guardian of Thomas and Richard, and
Thomas Staunton, his executor and first wife's brother, as guardian of Robert,
to make deductions for their 'reasonable support'. Guardians sometimes put in a
stiff bill. It looks as though the second Mrs Large and the first Mrs Large's
brother had quarrelled over their conflicting interests in the inheritance.

[3] No doubt a relative of the mercer Thomas Gedeney to whom a previous
William Caxton had been apprenticed in 1401.

[4]
> 'His prentice will I be seven year
> His science truly for to lere
> And with him I will dwell,'

says the Childe, who is rewarded for his virtue by becoming his master's
agent or 'attorney' ('and mine attorney that thou be'). The Wardens' Accounts

Ordinances included a special oath, 'so help you God and All Saints and by this Book', for the apprentice sent to foreign parts. The provisions seem a little more lenient than those of the general vow on enrolment, but show very clearly the repertoire of misdemeanours to be feared from the unfaithful servant. The apprentice was not expressly forbidden to buy, sell, or stand security on his own account, but must do so only by his master's 'consent, will, and agreement'. He must not take bribes in cash or kind to his master's detriment, betray mercers' business secrets, keep or share a private office or warehouse, nor 'play at dice, cards, tennis nor at any other disports or plays to any prejudice, hurt or harm of your said master'; and if he found anyone else committing any of these offences, he must promise to inform the Wardens.

Such a date as 1444 for Caxton's move to Bruges is probably the true meaning of another of his all too rare and ambiguous fragments of autobiography. 'I have continued by the space of thirty year for the most part in the countries of Brabant, Flanders, Holland and Zeeland,' he remarks in the prologue to his *Recuyell of the Histories of Troy*. Because he finished his translation of this work from the French original on 19 September 1471, it has usually been assumed that he calculated his thirty years back from that date. But it does not follow that he composed the prologue itself at the same time. It is more likely that he added it at the time of printing, which was not before 1474 and may have been as late as 1475; or, if he did write it earlier, that he updated his little sum when the text went to press, for his stay in the Low Countries was then still going on. If this interpretation is correct, Caxton's own figure of 'thirty years' in 1474 or 1475 confirms the likelihood that he went abroad towards 1444 or 1445.[1] A year or two later, in 1446 if he really was born in 1422, he would have reached his twenty-fourth birthday, his coming of age as a citizen, and the customary end of apprenticeship. Perhaps he continued for a year or so more as a hired employee, 'servant', or 'lowe' (*loué*) as it was called, for which the Mercers' Ordinances laid down yet another oath; or perhaps he felt he already had experience and capital enough to take the freedom of the Company and start in business by himself.

---

for 1431-2 show Thomas Staunton acting in Flanders as 'brother[-in-law] and attorney of Robert Large'.

[1] Blake points out truly that 'thirty' was often a round number. I think it more probable that in this particular context it was approximately accurate. However, our conclusion is much the same, that Caxton did not go to Flanders until the mid-1440s.

3

THE MERCHANT

CAXTON went abroad at a lucky moment. His apprenticeship
had begun in difficult times, when the cloth trade had slumped
and English merchants were barred from Flanders; for Philip the
Good, Duke of Burgundy, after fifteen years as England's ally
in the Hundred Years' War, made a separate peace with France at
the congress of Arras in 1435, and declared war on England. In 1436
Philip even besieged the wool-staple fortress towns of Calais and
Guines, but was chased away by Good Duke Humphrey and fled,
leaving behind (as Caxton's chronicler records) 'the great gun of
brass which was called Dijon and many other great guns and serpen-
tines'. After four years of hostilities Philip decided that neutrality
would be more profitable, and on 23 September 1439 a commercial
treaty or *intercursus* between England and Burgundy was signed for
three years, soon prolonged till 1447, and thereafter indefinitely
continued. The cloth trade with Flanders immediately improved.
Even the war with France was halted in 1445 by the marriage of
Henry VI and Margaret of Anjou. But that odious conflict would not
have lasted a hundred years if it had not been generally popular in
England, being immensely lucrative in loot and ransom-money,
gratifying to nationalist pride, and good for trade by opening
markets in the occupied territories of Normandy and Gascony. The
pro-war Good Duke Humphrey had to be liquidated in 1447; 'some
said he died for sorrow,' wrote the chronicler, 'some said he was
murdered between two featherbeds, others said that a hot spit was
put in his fundament, but how he died God knows to Whom is
nothing hid'.[1] The truce was broken in 1449, and within a year all
Normandy was lost, by 1453 Gascony and Bordeaux, everything
except Calais. The English nobility seized the opportunity to fight
one another instead, and began the Wars of the Roses. Lancastrians
looked to France for help, Yorkists favoured Burgundy; so the
mercers were strong Yorkists, for the sake of their Flanders trade,
which became more flourishing than ever now that competing
French markets were lost. London itself was vulnerable; in 1450 to
the rebellion of Jack Cade and his Kentishmen, a proto-Yorkist demo
that got out of hand and beheaded Lord Saye, treasurer of the realm, in
Cheapside; in 1458 to the anti-Lombard riots of 'the young men of the
mercery for the most part prentices'; in 1461 to Queen Margaret's

---

[1] Some thought he was starved to death, and this may be why going without
a meal was called 'dining with Duke Humphrey'. In Shakespeare (*Henry VI
part 2*, act 3, scene 2) he is strangled. Holinshed, followed by modern opinion,
believes he died of a 'palsy' (stroke) caused by the shock of his arrest.

pillaging army of northerners ('for the city of London dreaded sore to be robbed and despoiled if they had come'). Perhaps Caxton was well pleased to spend his thirty years in the Low Countries.

But even at Bruges one was not always safe. In May 1449 the Devon pirate Robert Winnington—armed, as was typical of those frantic times, with a commission to 'cleanse the sea of pirates'—captured the Flemish, Dutch and Hanseatic salt convoy of 110 ships, homeward bound from the Bay of Biscay, and brought it into Southampton Water for ransom. Philip the Good, who was just about to make ends meet by imposing a new salt-tax, retaliated as best he could, and 'the merchants of England being in Flanders were arrested in Bruges, Ypres and other places, and might not be delivered nor their debts discharged till they had made appointment for to pay for the amends and hurts of the ships, which were paid by the merchants of the Staple every penny'.

Our first document introducing Caxton as a freeman mercer must surely be somehow connected with this mishap, for it involves the sudden departure from Bruges of one of the visiting merchants of the Staple of Calais against whom the reprisals were chiefly directed. The stapler, John Granton, was arrested on the plea of the English merchant William Craes for debts of £60 and £35, but was allowed to leave after giving two other English merchants at Bruges, John Selle and William Caxton, as his sureties. Towards the end of 1449 the impatient Craes sued Selle and Caxton before the Sheriffs in the Town Council for a total of £110 including costs. The defendants pleaded that Granton would certainly pay Craes in due course, being solvent and rich, and disputed the costs. The Sheriffs gave judgement on 2 January 1450[1] that the defendants must pay £95 (evidently considering the alleged costs excessive), but that Craes must in turn hand over Granton's bond, give security for the same amount of £95, and repay this 'or more' if Granton should turn up and prove either that he did not really owe the money or that he had settled the debt already.[2] From the large sums involved it is generally supposed that this transaction reveals Caxton as a prosperous business man. Perhaps so, in spite of its apparently adverse outcome. Private surety jobbings of this kind were then a normal and essential means of credit dealing in the absence of international banking facilities. Craes no doubt foreclosed because in that particular year a stapler seemed a poor short-term risk; no doubt Granton eventually repaid Caxton and Selle with the usual interest. Interest existed then as now, but in

---

[1] The original document gives the year as 1449, but as the Flanders year began at Easter 2 January would be in 1450 in modern chronology. However, the antecedents of the case would all have occurred in 1449.

[2] Blades summarises the verdict wrongly, saying that (a) Selle and Caxton were made liable for the full claim of £110, (b) they were ordered to give security only (not cash), (c) that Craes, if Granton was able to disprove the debt, would have to pay a fine of 'double the sum claimed'. Duff, Plomer, Crotch and others have followed him, instead of troubling to read the document themselves.

various concealed forms, for to call it interest would have involved the deadly sin of usury; one might arrange to hand over a sum less than that to be ultimately repaid, or stipulate reimbursement in instalments at increasingly favourable rates of exchange, and so on.

In the autumn of 1452 Caxton was in London again, in time to take the livery of the Mercers' Company before the riding of the new Lord Mayor, Geoffrey Felding, on 29 October.[1] Admission to livery, the status-enhancing last stage in the full fledging of a mercer, entitled him to wear the Company's magnificent uniform on ceremonial occasions and to hold office. Perhaps it was partly absence abroad that had prevented Caxton from taking livery earlier; however, it was usual to wait several years after becoming a freeman, no doubt because it would have seemed over-forward for a very young man to seek this advancement. Among the other sixteen who paid livery fees in that year (1452-3) was Caxton's fellow apprentice Richard Bonifaunt, who was a still later joiner, for his enrolment fee as apprentice had been paid in 1430-1 and his issue fee in 1441-2. Others included Robert Cosyn and John Shelley, both of whom did business with Caxton a year later,[2] a fellow Kentishman William Pratt, whom Caxton called long after in his prologue to the *Book of Good Manners*, 1487, 'an honest man and a special friend of mine', 'my singular friend and of old knowledge', and William Pickering, whose kinsman John Pickering succeeded Caxton as Governor at Bruges in 1472. The seventeen new liverymen of that year all preferred to pay the £1 fee in three annual instalments of 6s 8d (except one, Henry Litelton, who settled in full). The payment entries in the Wardens' Accounts are here subdivided into groups, in which Caxton's name appears last of a trio with Emond Redeknape (no doubt a relative of the William Redeknape who became associated with Caxton in the 1460s) and Richard Burgh (who took a Company letter 'over the sea' —which usually meant to Flanders—in 1448-9 for a fee of 6s 8d). But this peculiarity is evidently due only to the accounting clerk, who recopied the entries from different pages or dockets. Possibly these three did apply on the same day, perhaps they even turned up together; but Blades and Crotch are not entitled to invent their little romance about all three travelling in company from Bruges specially for the purpose. The clerk has struck out the entry for Burgh and Caxton[3] with a marginal note: 'because these are among the debtors

---

[1] Blades gives the year of these incidents as 1453, and is followed by everyone. But the entries in the Wardens' Accounts are for the year 25 June 1452-24 June 1453, and mayors were elected and rode in October. The new mayor in 1452 was in fact Felding. He was a mercer, and a friend of the late Robert Large, whose third son and ultimate heir Richard Large had been his apprentice, issuing in 1442-3.

[2] Possibly this John Shelley (no doubt a member of the ubiquitous family which later produced the poet) was the same person as the John Selle (so spelt in the Bruges-French document) who acted as surety with Caxton in the Bruges affair of 1449-50?

[3] Blades and Crotch are mistaken in saying that Redeknape's name is deleted also.

at the end of the account'; and at the end he duly enters Burgh and Caxton as 'debtors for their first year for entry into the livery'. This trivial circumstance has been misunderstood in various ways. Blades thinks Caxton was so specially respected that the fee was remitted; Crotch imagines that the 6s 8d was a fine for dilatory payment; Blake apparently infers from the note of non-payment that Caxton's admission was thereby cancelled, and concludes: 'there is no evidence that William Caxton was ever admitted into the livery at all'. However, it has to be remembered that the Wardens' Accounts were not intended as an official register of admission and other incidents in the careers of Company members, but as a mere book-keeping ledger of receipts and expenses, compiled annually from other records which have long since disappeared. Clearly the clerk knew Burgh and Caxton *had* taken livery, but assumed wrongly that they had paid 6s 8d on the spot like the others, and discovered his mistake on checking back when he found his cash balance was 13s 4d short.

The next section in the accounts confirms that Caxton and Burgh were indeed received as liverymen, for they are both included in a list of thirty-six liveried mercers each fined 3s 4d for not riding in the procession of the new mayor Geoffrey Felding on 29 October 1452. If they had not been liverymen they would not have been under the obligation to ride. At that time the Mayor of London was elected each year on 13 October, was sworn in at the Guildhall on the 28th, and on the 29th rode to Westminster escorted by the livery companies of the City to present himself to the Lord Chancellor, a pageant which survives to this day as the Lord Mayor's Show. Oddly enough, this of 1452 was one of the last ridings through the streets for 400 years. Next year's Mayor, John Norman, says the chronicler, 'on the day that he should take his oath at Westminster went there by water with all the crafts, where aforetime the mayor, aldermen and all the crafts rode a-horseback, which was never used after, for since that time they have ever gone by water in barges'.[1]

At an unspecified date in 1453 Caxton experienced a little contre-temps, when goods being exported by himself and other English merchants were seized by the customs at Nieuport near Ostende, and then released on payment of duty. This venture was doubtless typical of the Englishmen's return trade in Flemish goods financed by their outward trade in woollen cloth, except for the astonishing wealth of the cargo, which included a bale of linen, a cask of furs, a cask of saffron, and 800 pelts of ermine. Perhaps Caxton shipped from Nieuport instead of Bruges in the vain hope of evading attention from the customs officials. Unfortunately the document in question seems subject to caution, pending full publication, for according to its discoverer Thielemans it concerns 'William Caxton governor of the English Nation', a post which Caxton did not acquire until 1462,

[1] The day was changed from 29 October to the modern date of 9 November in 1752, when the English calendar was put forward by eleven days, and the procession returned to land after the last water pageant in 1856.

whereas according to Blake it describes him as 'of the Staple of Calais', which Caxton is never called elsewhere. Staplers sometimes found it convenient to become mercers, but mercers had no incentive to become staplers. The possibility remains that the document may relate to Caxton's contemporary namesake, William Caston of Calais; but the activity described seems just like our Caxton, and it appears unlikely that the Calais man was a stapler or even a merchant of any kind.

Caxton again appears in London on 11 December 1453,[1] when he signed and sealed an extraordinary document, and attended in person next day to make formal recognition of it at the King's Chancery in Westminster. This was an assignment of his entire fortune to Robert Cosyn (the mercer who had taken livery with him the year before), and a certain John Rede. The agreement is so formulated as to leave no possible loophole: Caxton resigns all his property of any kind, movable or immovable, in England or abroad, in his own possession or deposited with others, whether goods, chattels, merchandise, cash, or money owed to him, for Cosyn and Rede to have sole and unhindered disposal now and for ever, and excludes himself and anyone else in perpetuity 'from any action, right, claim, title, or property'! Such a transaction was of course a legal fiction, of a kind not uncommon in the sophisticated vagaries of medieval law; indeed, Caxton himself in the same way, long after in 1487, took over the property of the mercer William Shore in association with two other persons. The assignor, no doubt, received in exchange a second document from the assignees certifying that the first was really null and void. Caxton's declaration keeps total silence as to its actual purpose. Most probably Caxton wished to secure his assets in London or Flanders from legal action during his absences from either place; or the assignment may have constituted security for a large or indefinite credit; or it may even have formed part of the procedure in a marriage contract involving mutual donation of property.

What is quite certain, however, is that Caxton was not in financial straits nor even seriously in debt to Cosyn. Only nine months later, on 7 September 1454, Cosyn testified before the Mayor and aldermen at the City Council that *he* owed Caxton the large sum of £290, to be repaid partly in woollen cloths[2] and partly in pewter vessels; and Cosyn also handed over to John Shelley (another of the mercers who took livery in Caxton's year), who was acting on Caxton's behalf, a consignment of cloth to the value of £72 as part payment on account. Evidently the situation has now entirely changed; indeed, it

---

[1] When 1453 was wrongly supposed to be the year of Caxton's livery and Felding's procession it was naturally supposed that this occurrence of 11 December 1453 belonged to the same visit to London; but now our new evidence suggests that Caxton returned to Bruges during the thirteen-month interval.

[2] Blake's reading '*de la vyse*' should probably be '*de l'assise*', i.e. cloth of assize or broadcloth of standard measurement.

seems likely that this incident is not directly connected with the one before or with the one after, except that in all three Caxton is in business relations with the same Robert Cosyn.

Next spring Caxton was again in London, when he gave evidence at the City Council meeting of 5 May 1455. John Neve, mercer, testified in the matter of a debt owed to himself by Cosyn that he (Neve) had previously been in partnership with Caxton at Ghent in a bargain concerning linen cloth for which he still owes Caxton £200. Then Caxton stepped forward to depose that Neve had stood security for him for the sum of '£80 and more', and that he had repaid Neve £36 of this as part of their Ghent bargain. He himself, Caxton added, was bound security for 1000 marks (£666 13s 4d) to 'certain persons'; and he undertakes, failing successful arbitration of his dispute with John Harowe,[1] to attend the Council 'within a year and a day' and prosecute Harowe to safeguard Cosyn from loss. These proceedings evidently do not relate to a single transaction, but to a whole chain of credit relations, which has to be unravelled in reverse —rather as in the folk tale of the old woman getting over the stile— so that in the end the first transaction, between Neve and Cosyn, will be settled last. Twenty years later, as we shall see, Caxton was still doing business as a mercer with Neve, although at that time he had already begun his career as a printer.[2]

The documents which we have now discussed are merely random archival survivors from a much greater volume of activity, but suffice to give a true sampling of Caxton's life in the 1450s. The mercer trading with Flanders bought woollen cloths and other wares at home in England, and shipped them to Bruges. He might sell them there on the spot to resident Flemish or visiting foreign merchants; but his chief business came from attending the four great seasonal fairs of the Netherlands—quaintly called Cold mart (in winter), Pask mart (at Easter), Synchon (St. John's Day) or Pinxster (Pentecost) mart (at midsummer or Whitsuntide), and Balm mart (St. Bavo's Day, 1 October)—at Bruges itself, Antwerp, Bergen-op-Zoom (which the English called Barrow), and sometimes Middelburg. With the proceeds he bought the manufactured goods and luxury articles of the Low Countries, and sold these on his return to London, so gaining a double profit on the double journey. Each bargain was normally based partly or wholly on interest-bearing credit, disguised as an interest-free loan, and was often made semi-public by declaration in a court of law or city council. English staplers from Calais regularly attended the fairs at Bruges and elsewhere, collecting last season's credit debts in cash from their Netherlands

---

[1] We have met Harowe twice already as Caxton's predecessor among Large's apprentices, and as the master from whom Richard Caxton issued in 1446-7.

[2] The relevant document is dated 28 January in a year which must be 1475. Blake's view that it belongs with the three documents of the 1450s is mistaken, as is his opinion that all four relate to one and the same 'lawsuit'. They concern, rather, separate incidents in Caxton's dealings over a period of many years with Cosyn and Neve.

customers, and transferring the money to London by negotiating
bills of exchange with English merchants, for whom this was a
useful source both of profit and of Flemish coin. Caxton's operations
fit perfectly into these patterns. We find him visiting London in the
autumns of 1452 and 1453 and the spring of 1455; he obtains both
cloth and pewter ware from Cosyn in London for export, and attends
the Ghent fair to buy linen goods with Neve for import; he obliges a
stapler at Bruges in 1449, suffering temporary hitches which will
no doubt end in profit; and he engages in multiple chains of credit
and loan operations in which the same persons, Cosyn, Neve, himself
and others, appear alternately as creditor and debtor. It would be
misleadingly anachronistic to regard these transactions—except
perhaps for the Bruges contretemps of 1449-50 and Caxton's threat
to Harowe in 1455 (which may, however, have been only legal
byplay)—as involving litigation in the modern sense between
hostile parties for the recovery of debt, or to suppose that they show
Caxton or his associates in financial difficulties. The apparent
'debts', so called in medieval parlance, are really normal, friendly,
and mutually satisfactory credit dealings between business associates;
the apparent lawsuits are more in the nature of contractual state-
ments made in public by agreeing parties.

Only a single document has been found mentioning Caxton during
the next seven years, from 1455 until 1462. Let us fill the interval with
a glimpse of Bruges at that time, the Merchant Adventurers and their
house, and the circumstances that led to Caxton's next advancement.

Bruges was then one of the great cities of Europe in wealth,
magnificence and culture. Pero Tafur, the travelling Catalan nobleman, saw it in the bad year of 1438, when Duke Philip the Good had
just punished the citizens for a rebellious attack on his sacred person,
and severed heads were stuck on poles all round the city. Even so,
Tafur thought Bruges superior even to Venice, because, as he
observed, all the merchants of the West thronged there, whereas at
Venice only the natives were allowed to trade. Among the chief
sights were the Waterhall, a two-storied covered dock and cloth
market in the form of a superb gothic hall, where barges were un-
loaded and reloaded in the centre of the city; the church of
St. Donatian,[1] and its cloisters where the booksellers and manuscript
illuminators had their stalls; the Prinsenhof or ducal palace, one of the
favourite residences in the 1440s and 1450s of Philip the Good, who
had quite forgiven the repentant survivors of his tremendous
revenge; the almost equally palatial townhouse of the great burgher
Louis de Gruthuyse, with a library rivalled only by the Duke's own;
the public hot baths, where tourists flocked to see men and women
disporting together stark naked ('which they take to be as honest as
church-going with us', said Tafur); and the great crane, worked

[1] Not Donatus (as Blades) nor Donatius (as Crotch), but Donatianus,
Bishop of Rheims (died A.D. 390), the patron saint of Bruges, whose relics
were translated to this church in the tenth century.

by men walking in a treadmill, and adorned (so that no one could mistake what it was called) with painted effigies of crane birds. In Bruges one could buy luxury goods from all Europe and beyond: figs, oranges, lemons, gerfalcons, pepper, wax, quicksilver, furs, carpets, monkeys, and parrots. The local silks, satins, jewellery, candles and illuminated manuscripts were the finest in the world. By the 1460s even printed books were on sale, imported down the Rhine.

Every mercantile nation in Europe had its own communal residence, governor, charter of privileges, place of worship, and in many cases a street named after itself.[1] The English lived, met, slept, and dined in a stately hall in English Street, very like contemporary halls in Oxford and Cambridge colleges, but with dormitories above and cellarage for their goods below. The rules laid down that everyone must be in by ten o'clock (nine in winter), keep no dog, much less a mistress, call no man 'false knave' (except one's own servant), and not play cards for more than sixpence a game, or get very drunk too often. But these rather monastic regulations were inspired, like those of a London club or an officers' mess, more by mutual male convenience than by inflexible moral rectitude. A merchant could take private lodgings elsewhere and do as he pleased, while still enjoying the amenities of the English Hostel. All travellers agreed on the beauty, friendliness and exquisite attire of the ladies of Bruges; a visiting Queen of France had remarked: 'I thought I was the only queen in Bruges, but now I see there are many there.' The Waterhall offered special facilities. Any woman who wished might stay the night there, said Tafur, and the merchants could 'bring any woman they chose and lie with her, on condition that no one should endeavour to see her, nor to know who she was'. The possibilities of relaxation at Bruges need to be remembered, not because we have the least reason for doubting the Printer's spotless purity, but because some scholars have wasted their sympathy on his enforced continence. It has even been suggested that he married towards 1470, at the age of nearly fifty, because he was 'no longer willing to sustain the celibacy imposed by the rules of the Company on merchants abroad'. That he did marry is certain, as we shall see, and that he did so as late as 1470 is quite likely; but he may well have taken a first wife (who later died like Robert Large's) towards 1450, whether in London or at Bruges, and even raised a family, without leaving any trace of this in the surviving documents.

The English merchants in the Low Countries had been an organised body ever since the thirteenth century. The Rue Anglaise at Bruges was so named in 1285; the Duke of Brabant granted charters in 1296

---

[1] Letts lists hostels for the Austrians, Biscayans, Castillians, English, Florentines, French, Genoese, Germans, Irish, Luccans, Portuguese, Scots, men of Smyrna, Spaniards, Turks, and Venetians, and streets named after the merchants of Lubeck, Bayonne, England, Scotland, Ireland, Florence, Gascony, Bordeaux, Denmark, Hamburg, Norway, Portugal, Venice, and Bilbao.

and 1305 permitting 'the merchants of the realm of England' at Antwerp to hold courts and elect their own Governor; the Count of Flanders gave similar privileges at Bruges in 1359; the English merchants abroad were recognised by Edward III in a charter of 1353, and those 'in the regions of Holland, Zeeland, Brabant and Flanders'[1] received from Henry IV in 1407 a comprehensive charter which was confirmed in 1413, 1430 and 1437. In the pragmatic medieval way this convenient unity covered a more complex diversity, for the merchants also owed allegiance not only to a governing body in London, but also to their own parent companies, whether in London or in provincial centres, including York, Hull, Scarborough, Newcastle-on-Tyne, and Norwich. From the beginning of the fifteenth century, however, the London mercers became increasingly predominant in the Low Countries trade. Their first aggrandisement was due to the channelling of English wool exports through the Staple of Calais, and the consequent rise in the mercers' exports of woollen cloth, which was not subject to the heavy duties paid to the Crown by the staplers. By the middle of the century cloth had overtaken wool as England's chief export. Paradoxically the mercers in Flanders benefited also from the mid-century trade depression caused by the Hanseatic League's monopolisation of Norwegian, Baltic and German markets, and by the loss of Normandy and Gascony in the final stage of the Hundred Years' War, which left the Low Countries as the only free outlet for English cloth in Northern Europe. By Caxton's time the English Nation at Bruges were mostly mercers, and their Governor was ordinarily a mercer. The controlling Court of Adventurers in London was held in the Mercers' Hall at the church of St. Thomas of Acon, and consisted mainly of mercers, with a few visiting wardens from the drapers, grocers, skinners, and haberdashers, who also had interests in the cloth trade. Even its minutes were kept in the same book as those of the Mercers' Company. The Court appointed the Governor at Bruges, directed him in matters of policy and when it saw fit (as it twice did in Caxton's lifetime) dismissed him. Even so, in the democratic oligarchy of the Mercers' Company, where even the four Wardens held their posts for only a year, the Governor was the most influential, authoritative, independent, permanent, and highly paid official. Towards the beginning of the 1460s Caxton himself was in the running for this grand appointment.

[1] It is interesting to find Caxton automatically using this familiar formula in his remark about his sojourn of 'thirty years for the most part in the countries of Brabant, Flanders, Holland and Zeeland'.

4

THE GOVERNOR

THIS new departure in Caxton's career, and the silence that preceded it, fit the currents of political and economic history in a manner that cannot be wholly coincidental. Caxton's lost seven years between 1455 and 1462, when we find no trace of any visit to London and only a single document concerning business abroad,[1] cover a period of mounting tension between Lancastrians and Yorkists, culminating in three years of savage civil war and a change of dynasty. The mercers favoured the claimant Richard, Duke of York, and his son Edward, Earl of March, as heirs to the anti-French, pro-Burgundian, mercantile, middle-class policies of Good Duke Humphrey. In July 1460 Caxton's acquaintance the mercer John Harowe led the London militia in person to capture the Tower from the besieged Lancastrian garrison, who had bombarded the Yorkist citizens with their incendiary guns. Soon this tragic hero fought again, at the Yorkist defeat of Wakefield on 30 December 1460, and was executed in the usual bloodbath of prisoners; his severed head was exhibited on York gates with the others, including Duke Richard's, which wore a paper crown in mockery.[2] Caxton must have felt a special twinge of recognition, when he printed twenty-two years later the name of 'John Harowe of London captain of the foot' in the chronicler's list of victims. During the war years 1459-62 the cloth trade to Flanders and wool exports to Calais slumped by a third to the lowest level since the crisis of 1436. So it is no wonder if, as the gap in documents suggests, Caxton avoided turbulent London, where Lancaster threatened pillage and York levied enormous 'loans', or took less part in diminishing trade, and turned his talents and ambitions towards administration. No doubt politics and economics also explain the strange case of Governor Overey.

The governor of the English merchants at Bruges during most of the critical period from 1456 to 1462 was, quite anomalously, not a London mercer but a fishmonger, William Overey, of a prominent Southampton family. Worse still, from the viewpoint of Yorkist mercers, Overey had been appointed, in defiance of the Adventurers' and Mercers' privilege of electing their own choice, directly by the King of England, who at that time was the Lancastrian Henry VI.

---

[1] The customs records of Middelburg show Caxton importing the largish quantity of 30 'thousands' (about 15 tons) of iron on 16 April 1460.

[2]   *'Off with his head and set it on York gates,*
   So York may overlook the town of York,'

says Shakespeare's tigerhearted Queen Margaret in *Henry VI part 3*, act 1, scene 4.

From these antecedents and the sequel it is clear that Overey's was a political appointment, welcome to non-mercers and Lancastrian fellow-travellers among the English merchants at Bruges, but unpopular with mercers and other Yorkists, who would be glad when the time came to see one of their own in his place. Still, no one could accuse Overey of want of patriotic zeal. On 20 December 1456 he persuaded the Bruges sheriffs to punish a Florentine merchant Jacques Strozzi for insulting the English, with a fine of payment for a best-quality mass in the English chapel and a load of peat to be distributed to the poor by Overey himself. At the Antwerp Whitsun fair in 1457 the master of the public weighing machine accidentally spilled a bag of valuable madder dye belonging to an Englishman, who called him a beggarly knave, and was told in return that he 'lied like a tailed Englishman'.[1] Overey led his countrymen in an exodus back to Bruges, boycotted the October fair, threw the only English merchant who disobeyed into prison,[2] and secured not only an apology but a renewal and extension of privileges (for which of course his intransigence had been a pretext) from the repentant Antwerpers. In 1458 Overey appears twice (14 May, 20 August) in lists of mostly Yorkist delegates to the Anglo-Burgundian conferences of that summer on questions including the Earl of Warwick's piracies and land-incursions from Calais.[3] Caxton himself was included, receiving his first diplomatic appointment, after being summoned to Calais for preliminary briefing; for Philip the Good

[1] 'Tailed Englishman' (*Anglois coué*) was a double-barrelled taunt, to be taken both literally and figuratively. Everyone knew that Englishmen sometimes had tails, as a punishment (so folk-legend said) for pelting St. Augustine of Canterbury with fishtails at Dorchester in Dorset, or alternatively at Rochester in Kent, about A.D. 600. As Blake points out, Caxton himself was particularly sensitive on this point as a Kentishman, for he omitted the fishy tale when he translated and printed *Mirror of the World*, and when it turned up again in the *Golden Legend*, which as a semi-sacred book he could not well censor, he added: 'it is said commonly that this befell at Strood in Kent, but blessed be God at this day is no such deformity'. Blake suggests this remark may mean that Caxton himself was born at Strood, a village near Rochester. But perhaps it is only a touch of Caxtonian humour, intended to locate the mishap in a part of Kent as remote as possible from his birthplace in the Weald. More seriously, however, the expression also signified a complaisant husband, and Yorkist rumour had it that the termagant Queen Margaret's son Edward Prince of Wales was 'a bastard gotten in adultery', and the mild King had remarked that the Prince 'must be the son of the Holy Ghost'. So 'tailed Englishman' was an insult specially galling to Lancastrians. On tails in Kent see also p. 4, note 2, above.

[2] He was a greenhorn named Richard Charrety, and was forgiven on the plea that he had never been abroad before.

[3] Warwick, the future Kingmaker, was made Captain of Calais in 1455 and Admiral of England in 1457, supplanting Edward's brother-in-law Exeter, who turned Lancastrian from indignation. Warwick interpreted his task of 'keeping the seas' to include victorious piracy against a Spanish and Genoan merchant fleet bound for Flanders, and a Lubeck salt fleet. He was supposed to be acting on behalf of the Lancastrian court, with whom he finally broke in October 1458; but he took the opportunity to make a secret pact of Yorkist-Burgundian friendship with Duke Philip the Good.

issued a safe-conduct from Calais to Bruges (necessary against both English and Flemish marauders on that uneasy frontier) on 1 August 1458 to Caxton and 'Antoine de la Tour' (evidently the Milanese diplomat Antonio della Torre, who at this time was in Yorkist service). Overey's participation was perhaps non-political and ex officio, being required because trade matters were involved; indeed, he was able to obtain favour for his services in the next reign. In October 1459, however, when the Yorkist leaders had fled to Ireland and Calais after their first setback at Ludlow, Overey was appointed to an exclusively Lancastrian mission sent ostensibly to obtain a prolongation of the Anglo-Burgundian commercial treaty, which was about to expire, but really to buy support against the Yorkists with trade concessions which could only be to the detriment of the English merchants. Overey had now dirtied his Lancastrian fingers, and become unacceptable to his countrymen. Next year the mercers used the beginning of Yorkist victory to depose him. In 1460 he was imprisoned for debts at Bruges by the mercer John Pickering,[1] who is named as Governor in Bruges archives of September 1460 and 26 July 1461, with the mercer Henry Bomsted as his lieutenant or deputy. In 1461, at the urgent plea of his friends the mercers, the new King Edward IV confirmed by act of Parliament the grants made to the Company under the three Lancastrian kings Henry IV, V, and VI (here declared to be 'kings of England in fact and not in right'), including the privilege of electing their own Governor overseas.

Still Overey was not beaten. He obtained letters patent dated 16 April 1462 from the forgetful King, reappointing himself, 'our most dear and wellbeloved subject William Overey', Governor as from 1 May 1462. This nomination, made over Pickering's head and in defiance of the mercers, was typical of Edward's policy of reconciliation by restoring Lancastrian office-holders to their posts. Overey must have written his own patent, or ensured that it included all the best bits from his petition. The text contains pointed allusions to the 'many discords, annoys and dissensions among the merchants', for which this measure will 'provide convenient remedy', the 'good, faithful and acceptable services' of Overey himself, and the fact that he must have 'all such wages and profits', besides a great deal more, as 'in 1458, when he held and exercised the said office of governor' (this was also the year when he had made himself useful to Warwick). 'For such is our pleasure and so will we have it, notwithstanding any letters falsely crept in, obtained, or to be obtained contrary hereunto', the King is made to say, evidently referring not only to Pickering's coup in 1460 but to countermoves being prepared

---

[1] No doubt a relative of the John Pickering who was Governor at Bruges in 1445, when he made a treaty with Bergen-op-Zoom, and died in 1448, and also of William Pickering, mercer, executor in 1447 under the will of Robert Large's associate the Kentish mercer William Milreth, who died in 1445, and of the other William Pickering who took livery with Caxton in 1452.

by the mercers at that very moment; and if any dare disobey, 'we will see them so punished without redemption, that they shall be an example to all rebellious persons'.

But Overey's triumph from this pulverising patent was brief, for the mercers knew a trick worth two of that. Incriminating evidence was brought in from overseas, in the shape of a certificate dated 2 June 1462 from the town authorities of Antwerp, affirming that Overey had taken a bribe of £58 'to be friendly unto their town, and that therefore he omitted certain articles of privileges'. The real grievance against Overey, who in fact had acted with great firmness in the 'tailed Englishman' affair and obtained many concessions from Antwerp, was that he was a Lancastrian fishmonger and not a Yorkist mercer; still, £58 *was* a thumping large bribe. The Mercers' Company's annual general meeting on 25 June voted his dismissal. Meanwhile the good King had cancelled his patent of 16 April to Overey on 15 June, after issuing on 28 May 1462 a new one reaffirming the Adventurers' right to elect their own Governor.[1] Patents were not given for nothing. This one cost £57 10s,[2] which the Mercers advanced, but debited in their accounts to 'the fellowship beyond the sea for the suit of their privilege', until it was paid two years later. Overey was not ruined. He served King Edward in diplomatic missions in 1462, 1463, 1469 (when Caxton himself was one of his colleagues) and 1471, and became mayor of his native Southampton in the 1470s.[3] What became of Pickering? He is not named as governor after September 1461, and probably his governorship was regarded as terminated by Overey's patent and the Adventurers' counter-patent. He had a second time round as Governor from 1471, when he succeeded Caxton, till 1497, the year before he died. In 1483 Pickering was threatened with dismissal for disrespect to the wardens of the mercers, and made to beg pardon on his knees. In 1486 he was replaced, first by a lawyer and then by a draper, but soon reinstated. Perhaps nasty temper already stood in his way in 1462.

Overey's successor as Governor at Bruges was our William Caxton. The scanty minutes of the London Court of Adventurers do not happen to mention Caxton as acting in his new capacity until the summer of 1463, and do not name him in so many words as 'governor beyond the sea' until August 1465. The real date of

---

[1] Blades, and Hakluyt before him, seem to have created confusion among modern economic historians by citing Overey's patent of 16 April 1462 (which of course is *against* the Mercers) in mistake for the real 'Large Charter' of 28 May 1462, which superseded Overey's and was long treasured in the Adventurers' archives, but remains unknown and unpublished.

[2] So Carus-Wilson (1967), p. 153. Blades (1861) I, pp. 15, 89, reads the sum as £47 0s 10d.

[3] Crotch's suggestion that the mayor is another person of the same name rests on his identification of Overey with a Guillaume Aubriet, who received a safeconduct from France in 1429, and would be too old for the job in the 1470s; however, the context surely suggests that Aubriet was a Frenchman and nothing to do with our Overey, who first appears in 1456.

Caxton's appointment has remained almost unknown, although the relevant document was published in 1928. In fact the town archives of Middelburg show that he gave his good offices as Governor (*'meester van der Ingelscher nacien'*) on 12 August 1462 in an appeal against the detention of the good ship *Beryte* in the Thames at Gravesend. So his appointment as Governor must have followed immediately upon Overey's dismissal in June 1462.

Perhaps we may deduce Caxton's powers and emoluments from Overey's patent, as Blades and Crotch did, but with a little caution, as Overey was evidently spreading his butter as thick as he dared. If Caxton was granted the same, then he was empowered to give justice, in all cases not involving life or limb, in a court composed of himself and twelve merchants of his own choice, with six sergeants to carry out his fell arrests. He could appoint his own minor officials, such as 'correctors' or 'brokers' to witness bargains, 'alnagers' to measure cloth, and packers to pack or unpack bales and prevent smuggling. He received twopence for every pack sealed for export, and a penny for each bargain witnessed, together with other 'accustomed dues'. Probably he also took a regular salary, as did the mercer John Wareyn, who was governor in 1421-30 with a wage of 200 nobles, or £66 13s 4d. So now, subject only to the overlordship of the everchanging annual Wardens, Caxton was the most important Mercer and Adventurer of them all.

Perhaps Caxton gave up private trade during his governorship. A decision of the town council of Middelburg on 16 July 1462 shows that William Caxton in partnership with other English merchants unnamed had chartered a ship from Pieter Willemszone to carry goods to London, but refused payment because part of the cargo was lost by shipwreck on a sandbank. The court decided that the English must pay the agreed charter fee, but could deduct their costs for salvage and re-transport. This is only the second document since 1455 which shows Caxton engaged in personal trade, and it is the last in his mercantile career. The original transaction must have taken place several months before he became Governor. This apparent abstention from personal business may well be real, and would be due to various causes, such as Caxton's lack of leisure, change of vocation, ampleness of perquisites from his office, and the persistent and unexpected trade difficulties of the 1460s.

A new commercial sun of York might have seemed to rise for the Adventurers in Flanders, now that they had a mercer Governor and a King of their own favour. Yet the 1460s were a decade of insecurity, when civil war never quite subsided, and a single lost battle might at any moment restore the House of Lancaster. In Flanders Duke Philip the Good leaned over towards the new Yorkist dynasty as far as his sympathetic neutralism would permit. He welcomed King Edward's little brothers George (aged eleven, soon to be Clarence) and Richard (aged eight, soon to be Gloucester) as refugees at Bruges in April 1461, after the tableturning Lancastrian success of

St. Albans on 17 February.[1] The city of Bruges gave the boys a farewell banquet, and this was perhaps Caxton's first contact with specimens of the royalty whose names he later so delighted to drop in print. Philip even sent a Burgundian detachment to fight for Edward at his conclusive victory of Towton on 29 March. But even Philip had to keep in with the Lancastrians. The next refugee at Bruges was the charming young Lancastrian magnate Henry Beaufort, 3rd Duke of Somerset, victor at Wakefield and fugitive from Towton, who was supposed to be the lover of termagant Queen Margaret, like his father before him,[2] and was certainly the darling friend of Philip's son the Count of Charolais, the future Charles the Bold. The English merchants showed their Yorkist sentiments by rioting in the streets of Bruges in March 1462 against the handsome duke and his unpopular retinue of northcountry men-at-arms.[3] The exiled Margaret herself visited Bruges in August 1463, and was royally welcomed by Charolais, at Philip's reluctant order, with money and a gala tournament, and an escort of archers to protect her from the English. The Duke of Exeter, Edward's doggedly Lancastrian brother-in-law, begged from door to door at Bruges in 1466, until Philip gave him a little pension. There were always a few prominent Lancastrians kept in cold storage at the court of Flanders, just in case. At any moment the Lancastrians might return to the palace at Westminster, and Edward himself would be a fugitive at Bruges, as indeed was to happen in 1470-1. Meanwhile Edward himself seemed surprisingly unforthcoming towards his friends in Flanders. The new King was determined to carry out the Yorkist policy of encouraging commerce, and even traded personally, tax-free of course, in wool and cloth, as did Warwick in wine. For the London merchants, however, improvement of commerce signified more privileges for themselves and fewer for foreigners, even to the detriment of their fellowcountrymen in Flanders. So Caxton, as it turned out, had been brought in to replace the Lancastrian Overey as a Yorkist caretaker during a decade of chauvinist protectionism, restrictions, and reprisals.

Caxton's duties often obliged him to enforce measures imposed by pressure groups at home in England against the interests of his fellow

[1] Shakespeare bends time and space with Proustian effect in *Henry VI, part 3*, act 2, when he makes Clarence and Gloucester fight as grown warriors at Towton, on a day when in fact they were child fugitives on their way to Flanders.

[2] Somerset's father Edmund, 2nd Duke, slain at the first Battle of St. Albans in 1455, was thought the most likely candidate for the paternity of Edward, Prince of Wales, who was born during Henry VI's insanity in 1453.

[3] Somerset charmed Charolais in 1460, when he made the outrageous offer to surrender Guines to him, the key to Warwick's Calais, but Duke Philip vetoed the idea. Somerset next charmed Edward IV by yielding Bamborough to him on 26 December 1462, and became the King's boon companion and even bedfellow. But he changed his mind and turned Lancastrian again, until he was routed and beheaded after the Battle of Hexham on 15 May 1464, aged twenty-eight.

Adventurers in Flanders. Edward's second Parliament of May–June 1463 protected English manufacturers by clapping an embargo on imports of a variety of luxury goods from Flanders,[1] 'brought from beyond the sea as well by merchant strangers as by denizens', including silks, gloves, purses, hats, daggers, razors, tennis balls, dice, and playing cards. This restriction cannot have been agreeable to the Adventurers, who depended for half their livelihood on their return cargoes of Flanders wares; even so, Caxton's first recorded act as Governor is embodied in a minute of that summer of 1463 in the Mercers' Acts of Court on 'a letter sent from William Caxton and the fellowship beyond the sea directed to my Lord Chancellor [then George Neville, Warwick's brother] as for the best restraining[2] of buying of ware at Bruges'. The wardens Hugh Wyche and John Stockton in person took it to the Privy Council at Westminster by boat up the Thames ('boathire, sixpence', noted the account clerk). Caxton was instructed to ensure that 'what person of the fellowship be found guilty in buying of ware at Bruges shall pay the fine thereof made at the discretion of the Wardens'.

Another recurrent problem of the 1460s was the renewal of the commercial treaty or 'intercourse' with Burgundy, which had been allowed to fall into abeyance since the last Lancastrian mission of October 1459, in which Overey had served. A first prolongation of one year was obtained in October 1461 at Valenciennes, where Philip grandly treated the envoys to 'baths equipped with everything needful for the joys of Venus, so that each man was free to choose whatever he best fancied'. Delegations which again included Overey obtained two further annual renewals in 1462 and 1463.[3] But in 1464 Caxton himself was at last appointed as the merchants' representative.

[1] His first Parliament in November–December 1461 had made a similar petition, which Edward thwarted by granting it 'subject to the consent of the Duke of Burgundy', which of course was not to be had.

[2] Blades misreads as 'reynyng'.

[3] In 1462 Overey was appointed to the delegation in September, but was apparently replaced by Louis Gallet, a prominent French Yorkist diplomat. The agreement of October 1463 was made during the diet of Saint-Omer, which Philip had narrowly prevented Queen Margaret from gatecrashing during her visit to Bruges. It seems likely that Overey was left out of subsequent missions, until 1469, for fear of his exerting undue Lancastrian influence.

5

THE DIPLOMAT

THE Anglo-Burgundian trade negotiations of 1464 were expected to form the last and least important stage of a summit meeting at Saint-Omer between Philip the Good, Louis XI of France, and Warwick. But the summit (as has sometimes occurred in modern times) broke up before it began. Kingmaking Warwick had concocted a royal marriage between his protégé Edward IV and Louis's wife's sister Bona of Savoy, partly in the hope of an anti-Burgundian alliance with France, partly to consolidate his power over Edward by imposing a consort of his own choice. But on 14 October, when Philip and Louis had been kept waiting and fuming for Warwick ever since June, the astounding news arrived, that Edward could not get away to marry Bona, because his wife would not let him. On 14 September the young King of England (Edward was still only twenty-two) had confessed to his Council that on 1 May he had secretly married Elizabeth Woodville, a ladylike commoner and widow aged twenty-seven, whose enormous and ambitious family had been staunch Lancastrians until they turned Yorkist after Towton, only three years before. Much as he loved his excellent Queen, it was the united Woodvilles rather than herself that Edward had deliberately resolved to wed, as a readymade King's party, devoted to himself, agreeable to reconciled Lancastrians and the middle classes, and a counterpoise to Warwick, who was their deadly foe.[1] Warwick was furious, seeing himself flouted, duped, permanently maimed in power, and made the laughing-stock of Europe. Louis, though out-tricked for once, was delighted to know that Warwick was henceforth his friend and Edward's enemy. Philip took the opportunity to quarrel with Louis and make friends with his own estranged son and heir Charolais, the future Charles the Bold. Edward hastened to give jobs, titles, and noble marriages to all the Woodvilles.

Amid these preoccupations it was no wonder that the King forgot to appoint envoys for the 1464 trade delegation until mid-October, when the Mayor and aldermen made so bold as to remind him. He signed a belated commission on 20 October, nominating 'our well-beloved subjects Richard Whetehill, Knight, and William Caxton'.

[1] Warwick kidnapped the two chief Woodvilles, Richard Earl Rivers and his son Anthony Lord Scales, at Sandwich in January 1460, and gave them a formidable scolding at Calais: Rivers' father 'was but a squire' and himself 'a knave's son, that he should be so rude to call him and these other lords traitors, for they should be found the King's true liegemen when he should be found a traitor'. The full irony of these words was revealed in 1469, when Warwick himself turned his coat again and beheaded Rivers senior for loyalty to Edward.

Whetehill was a Warwick man, who had served the Kingmaker as lieutenant at Guines, taken lavish presents and promises from Louis, and dined with Louis and Bona in that matchmaking summer of dupes. So Caxton was in with the Yorkist establishment at last, although, as it turned out, on the wrong side. They obtained a year's renewal of the 'intercourse', signed by Philip at Lille on 28 October 1464, but preceded two days before by another edict which made their success illusory. Philip's subjects had complained not only of Edward's ban on the importation of Burgundian manufactures, but of the unfair competition of English cloth, the excessive price of English wool from Calais, and the staplers' insistence, in obedience to another act of the 1463 Parliament, on payment in gold or silver coin, 'by which all the bullion of our countries goes to the kingdom of England'. So Philip in reprisal had issued on 26 October a decree declaring: 'It pleases us to ban the cloth and tissues of England entirely from all our countries.'

Officially the main items of trade between England and Burgundy were now prohibited on both sides of the Channel. Caxton had failed in his first task as a diplomat, though through no fault of his own, for the main decisions had been taken far above his head; but he could still act vigorously as Governor of the Adventurers. Taking a leaf from Overey's book, he ordered the English merchants out of Bruges and led them in exodus to Utrecht. Possibly the delighted Utrechters had Philip's tacit consent to welcome the English. In theory the Principality of Utrecht, between the Rhine and the Zuyder Zee, lay just outside his dominions; but their bishop–ruler, David, Bastard of Burgundy, was one of Philip's innumerable illegitimate sons,[1] and had been forcibly imposed on them by Philip himself in 1456. On 21 November 1464 the town council of Utrecht granted 'a year's safeconduct to William Caxton, governor, and the merchants of the English nation for their persons and goods', followed on 27 December by a licence to hold fairs. A tax of half a silver penny was to be paid on each piece of cloth by both buyer and seller. The English held a winter fair at Utrecht in January and February 1465, and spring and summer fairs from April to July. Apparently a special invitation was given to the merchants of Cologne, a city which would thenceforth be important in Caxton's career.[2]

Thanks to Caxton's exodus to Utrecht the interruption in English cloth exports after Philip's ban was brief. Customs records show that during the first nine months from December 1464 shipments from London fell dramatically to a mere 776 cloths, but then regained the

[1] Philip's bastards were perhaps as many as twenty-six, but, as Professor Vaughan wittily remarks, 'the severe pruning of historical scholarship might reduce the numbers to fifteen'.

[2] A copy of the invitation to the Utrecht fairs exists in the archives of Cologne. The two cities were conveniently linked by river transport on the Rhine and the then still navigable channel of the Old Rhine, on which Utrecht was situated. Utrecht and Cologne were also the furthest western centres of the Hanseatic merchants.

usual annual total of about 8000. Meanwhile the merchants perhaps eked out their stocks less scrupulously than they ought, for on 16 August 1465 the London Court of Adventurers met to consider 'evil measure of cloth and lawn'. William Redeknape and others had complained of short measure 'in all white cloth and brown cloth as in broad of the same, and in like wise in lawn nyvell [snow-white] and umple [fine linen].'[1] On 4 September Henry Bomsted (Pickering's former lieutenant) was sent with a letter instructing 'William Caxton governor beyond the sea as well for reformation of the precedents as for other &c'.

Anglo-Burgundian relations continued to worsen. Edward bided his time, yielded to the chauvinist pressure of home-based merchants, and left foreign affairs as a consolation prize to the pro-French and anti-Burgundian Warwick. The Parliament of January 1465 retorted to Philip's edict by extending the ban on importing Burgundian merchandise to cover everything except food; and the King's searchers in that year seized tennis-balls, hats, playing-cards and featherbeds to their hearts' content.[2] Once again the Saint-Omer diet had to be cancelled, for Charolais and Louis spent the summer of 1465 fighting one another in the strange War of the Public Weal. But Edward remembered the 'intercourse' this time, and graciously urged the Mayor and aldermen to name their own envoy; they in turn, much embarrassed, consulted the Court of Adventurers; and the Court replied on 17 October 1465, declaring that 'it is not our part here in the City to take upon us a matter of so great weight', and advising the Mayor to ask the King 'in the most pleasant wise that he can' to make his own choice. Meanwhile the Wardens wrote to Caxton, informing him of all this, and instructing him 'in as goodly haste as ye can' to 'labour for a mean by the which your persons and goods may be in surety for a reasonable time', until 'there come writing from the King to the Duke or else from the Duke to the King for the prorogation of the same'. Sure enough, Edward sent his herald Rougecroix Poursuivant to Philip, and the now almost meaningless intercourse was patched up once again.

No solution was found in 1466. An Anglo-Burgundian conference met at Saint-Omer in April, but made no agreement on trade. No better could be expected, for the chief delegates were the usual Warwick men (including Wenlock and Whetehill) headed by Warwick himself, who took an instant and fully reciprocated dislike to Charolais, and soon left for Calais to intrigue with Louis. Next month Warwick in person wrote a stern letter to Caxton, ordering him to enforce the 1465 ban on the purchase of Burgundian goods,

[1] Blades misreads as 'purple'.

[2] The London Court of Adventurers passed an outspoken resolution on 21 March 1465, deploring the danger that the King himself might evade this Act by granting special import licences 'to strangers and denizens too, to bring in at their leisure such as shall please his Highness to accept, which is likely to grow to a general injury if it so happened, which God forbid'; but they prudently decided to do no more about it.

which the English merchants had no doubt evaded as often as they could. Caxton wrote in perturbation to the London Court on 27 May 1466 from Bruges, enclosing a copy of Warwick's letter, asking for instructions, and warning of possible 'jeopardy'. The Wardens received his message on 3 June and replied immediately: 'Right trusty Sir,' they wrote, they 'willed in no kind the said act to be broken nor hurt by none of our fellowship'; Caxton must find and punish offenders 'quickly if any such be as God forbid';[1] his letters from Saint-Omer had not yet arrived, 'and as for any jeopardy that should fall, ye shall understand it there sooner than we here, and if we know of any ye shall have writing'. This answer was carried on 4 June by one Simon Preste. We may note with interest that a letter from Bruges took only seven days to reach London and be answered; that Caxton had evidently attended the Saint-Omer conference, although his name does not appear in the official list of envoys, and hence that his diplomatic activities during the decade may well have been even more extensive than the surviving documents reveal; and that he had not severed his connection with Bruges, despite the exodus to Utrecht.

Next year the deadlock between England and Burgundy began to break at last. Philip the Good died at Bruges on 15 June 1467, aged seventy-one, and Charolais became his own master as Charles the Bold, Duke of Burgundy. Ever since the death of his second wife Isabel de Bourbon, on 25 September 1465, Charles had been an eligible widower, and dallied with the idea of a political marriage with Edward's sister Margaret of York—much against the grain, for he preferred Louis to Edward, Lancastrians to Yorkists, and men to women. He had exchanged a secret treaty of friendship with Edward on 23 October 1466, and received an embassy at Bruges in April 1467 to negotiate for a package deal on the marriage and the trade war. This time, as he did when he meant business, Edward sent his own men, not Warwick's, including Richard Beauchamp Bishop of Salisbury, the King's physician and secretary William Hatcliff, Sir Thomas Vaughan, and the rising Yorkist churchman John Russell, a future patron of Caxton. No doubt Caxton took part in the discussions, and perhaps he was even present on 17 April when Russell bought a copy at Bruges of Fust and Schoeffer's second edition of Cicero, *De officiis*, printed at Mainz the year before on 4 February 1466, which still exists in Cambridge University Library and is the first surviving printed book known to have reached England.[2]

Meanwhile Edward again allowed the deluded Warwick to conduct

---

[1] Perhaps it was for an offence of this kind that Caxton had a certain Hugh Dommeloe imprisoned in fetters at Antwerp on 5 December 1465.

[2] Russell bought another work of Cicero at Bruges on the same day; but this was not another copy of the printed *De officiis*, as Duff said, but a Venetian manuscript of *Epistolae ad familiares*, which is now Lambeth Palace Library MS. 765. He bought the manuscript, and probably the printed book as well, at one of the booksellers' shops in the cloisters of St. Donatian's church. A record exists of 'two printed Bibles' imported from Cologne in 1466-7 for John Tiptoft, Earl of Worcester.

counter-negotiations with Louis. In May 1466 Louis had offered the wide choice of four desirable French-sponsored husbands for Margaret of York, including Galeazzo Sforza Duke of Milan, and Philibert, heir of the Duke of Savoy, with dowry and expenses paid, and a pension of 40,000 crowns for Edward. In April 1467 Louis proposed a scheme of exceptional outrageousness, even for him. He would arrange French marriages for Margaret and Gloucester; Clarence would marry Warwick's daughter Isabel Neville; Louis and Edward would unite to destroy Charles and share his dominions, giving Holland and Brabant to Clarence. This done, as Louis further promised in July, he would abolish the fairs at Antwerp and elsewhere, and open new markets for English merchants in France. By way of a contingency plan Louis invited Margaret of Anjou to his court; if the Burgundian marriage went through, he would help Warwick to rebel and make Clarence king, or restore Henry VI!

At this moment, when the peril from France became unmistakable and his neutralist father was just dead, the reluctant bridegroom began to yield. Charles renewed his treaty of amity with Edward on 17 July 1467; on 28 September Edward made the indispensable unilateral concession of cancelling the 1463 and 1465 trade bans; Margaret appeared before the great council on the 30th to declare her consent to the marriage. Edward sent his brother-in-law and right-hand man, Anthony Woodville Lord Scales, to head the final negotiations at Bruges; perhaps Caxton then met this future patron for the first time. The marriage, and a treaty of mercantile intercourse to last thirty years, were agreed at last in February 1468. Charles had driven a hard bargain, with unpleasant implications for the English merchants. The ban on English cloth remained, pending discussion at a diet to be held at Bruges after the wedding. The price of Charles's hand was a dowry of 200,000 gold crowns (about £40,000) 50,000 down on the wedding day, and the rest in yearly instalments. The marriage was thrice postponed while Edward struggled to raise the huge down payment, which was loaned at last by the longsuffering staplers on the security of bonds from eighty-six of the wealthiest London merchants, mostly mercers. The English at Bruges were obliged by the terms of the marriage settlement to give surety for later instalments, so that Charles would be entitled to distrain on them in case of nonpayment.

Margaret and Charles were married on 3 July 1468 at Damme, the outer port of Bruges. The English merchants, no doubt with Governor Caxton at their head, escorted the happy couple through the streets next day (it poured with rain, of course) under the banner of St. George. As they came in two eminent Lancastrian refugee dukes, Exeter and Edmund Somerset (brother and successor of Charles's beheaded friend) slipped out, but were recalled by Charles when the festivities were over. The nine days of revels included fireworks, tournaments, pageants, and fountains running wine; young John Paston wrote home: 'as to the Duke's court, I heard never of none like to it, save King Arthur's court'. Scales, Russell, Wenlock, Hatcliff and Vaughan were among Caxton's acquaintances in Margaret's train. Edward had also sent his favourite court jesters, John the Wise[1] and Richard the Lovelorn, in case Margaret should need cause for mirth. Charles chivalrously kissed his bride in public, 'and all the ladies that were English, but no other'. Margaret was to see little of him, and they had no children. 'Proud, irascible, and somewhat bestial', Louis had called him, and there is reason to suppose that Charles the Bold (like many other warlike rulers, such as Frederick the Great, or our own Richard Lionheart and William III) was homosexual. For the second time Edward had duped and flouted Warwick with a political matrimony; his Woodville marriage had made him master at home, and the Burgundian marriage ruined Warwick's pro-French foreign policy; now he had deprived Warwick of all power, except for open rebellion.

The Bruges diet on trade, after being postponed from 20 March to 10 May, and then to 28 June to coincide with the wedding, broke up on 18 July 1468 with no decision except to summon an advisory conference of merchants at Antwerp on 15 September. This time, when the King again asked them to name their own delegates, the London Adventurers had no excuse for bashfulness, and on 9 September they appointed William Redeknape (who had complained of short measure in 1465), John Pickering (who had served as interim governor in 1460-1), and William Caxton. Apparently Redeknape and Pickering were then still in London, for on the 28th the Court awarded them £40, to be borrowed from the Convoy fund (intended to protect merchant shipping from Channel pirates), 'toward their costs and charges for the embassy of the enlarging of woollen cloth in the Duke of Burgundy's lands'. However, the Antwerp meeting for traders only was soon cancelled and replaced by a joint congress of diplomats and merchants at Bruges, which was planned for 20 January 1469 and then postponed until 12 May, and again until 1 June. On 1 May Edward commissioned Russell, Wenlock, Hatcliff and others, together with seventeen merchants headed by Caxton himself and the mayor of the Calais Staple, John Prout; on 4 May he empowered Wenlock, Caxton and Prout who were already overseas to act alone if the rest were delayed by contrary winds; and on 23 May he appointed further specialist delegates to deal with monetary problems, including Caxton's future patron Hugh Bryce,[2]

---

[1] John le Saige (his real name was Woodhouse) once entered the royal presence with a fenman's leaping-pole and wearing thighboots. When Edward asked him why, 'Upon my faith, Sire,' replied the wise fool, 'I have passed through many counties of your realm, and in places the Rivers be so high that I could hardly scape through them, but as I was fain to search with this long staff.' Rivers, of course, was the earldom title of the Woodvilles. Edward, according to the chronicler, much enjoyed this jest ('made thereof a disport').

[2] Bryce was a London goldsmith and alderman of Irish origin, who had helped Edward to raise the down payment on Margaret's marriage settlement by borrowing £1676 from his City friends on the security of the King's jewels.

and none other than his old rival William Overey. Several documents show that during the negotiations Caxton travelled far and wide over the Netherlands. He was already away from Bruges on 12 May 1469, when the sheriffs accepted his written arbitration ('because the said William Caxton was unavoidably absent from the said city') in a case between a Flemish shipowner and a Genoese merchant of illustrious name, Jacques Doria. The archives of Middelburg record the expenses of messengers sent to Caxton at Bruges and Antwerp, and the city of Bruges paid for his wine at Ypres in August.

Only the monetary conference, which reached agreement on 23 August for exchange rates of English and Netherlands coins, was successful. Charles still evaded any decision on the ban against English cloth. The delegates were also instructed to negotiate in another intractable problem of the 1460s, the quarrel with the Hanseatic League. The Hanse possessed special privileges in London, including their own premises in the Steelyard (where Cannon Street Station is now), but for many years had refused reciprocity by excluding English traders from the Baltic and German ports. In response to pressure from London merchants, Edward had confirmed the Hanse privileges only for yearly or two-yearly periods (in 1461, 1463, 1465, 1466), made repeated efforts to call the Hanse to conference, and in July 1468 closed the Steelyard, arrested the Hanseatic merchants, and fined them £20,000.[1] Although Charles the Bold tried to mediate in favour of the Hanse, the Bruges meetings of June 1469 ended in a complete break; the Hanse recalled their merchants from England, banned English cloth, and declared naval war. But Cologne, westernmost and nearest to England of the greater Hanse towns, seceded from the rest, and was allowed to retain its privileges in London, with sole possession of the Steelyard. Caxton himself took part as intermediary in the negotiations; on 28 January 1469 the Mayor and aldermen of London received a letter from him 'written on behalf of the merchants of the Hanse of Germany', and ordered a full session in the Mercers' hall 'at the church of St. Thomas of Acon' to consider it. Caxton's links with his Hanse rivals and colleagues must have been numerous during

---

He was keeper of the Exchange and a governor of the Mint. When the Parliament of 1478 made a rather clever law to solve the Irish problem by sending all Irishmen back home or taxing them for the maintenance of order in Ireland if they refused to go, Edward made a special exception for Hugh Bryce. Bryce was not a mercer, as Blades and all after him have said, nor a Kentishman, as Blades called him in the mistaken belief that his manor of Jenkins was in Kent (in fact it was in Essex at Barking), nor the same person as the Thomas Bryce who was fined along with Caxton for not riding with Mayor Felding in 1452, with whom Crotch confuses him.

[1] The pretext for this action, known as the Verdict, was reprisal for the seizure of four ships, in which Warwick had shares, in the entrance to the Baltic by the King of Denmark in June 1468 as compensation for raids on Iceland by Bristol fishermen. The Verdict was forced by Warwick elements in the Council, and was the Kingmaker's last attempt to influence foreign policy before his break with Edward.

the last twenty-five years, for Bruges and Utrecht were the chief Netherlands headquarters of the Hanseatic merchants, especially those of Cologne; but the new special relationships with Cologne was soon to prove momentous in his career. Edward ratified the Cologners' exclusive privileges on 18 July 1469. This was to be his last act as a free monarch for nearly two years.

During a lull in this busy year Caxton had found time to begin his first translation, the *Recuyell of the Histories of Troy*, at Bruges on 1 March 1469,[1] wishing, so he says, 'to eschew sloth and idleness, which is mother and nourisher of vices, and to put myself into virtuous occupation and business'. His pretext is a stock convention of medieval writers, no doubt intended with double irony by this most industrious of men in that most strenuous of years. However, when he further alleges that he began work 'having no great charge of occupation', we may take him more literally, for March 1469 is the very time when Caxton was detained unwillingly idle at Bruges, waiting for a decision from England on the opening of the still postponed conference. After completing 'a five or six quires' he laid his task aside, as he modestly explains, 'when I remembered myself of my simpleness and unperfectness that I had in both languages, that is to wit in French and in English, for in France was I never, and was born and learned mine English in Kent in the Weald'; but here, certainly, he is again using a conventional 'humility formula', for Caxton was, and knew he was, surpassed by few Englishmen of his time as a master of French and of English prose. The true cause was no doubt the return of urgent business when the King on 1 and 4 May 1469 appointed him as delegate with power to act alone; and by 12 May, as we have seen, Caxton was already 'unavoidably absent' from Bruges. So, he says, 'I fell in despair of this work and purposed no more to have continued therein, and those quires laid apart, and in two year after laboured no more in this work'. Here again we may believe him literally, for there is no sign of resumption until early in 1471, and the two intervening years were the most precarious in all his career.

It was no wonder that the Bruges conference of 1469 broke up in August without reaching a decision on the cloth embargo. Suddenly the Wars of the Roses had erupted again, England was in a state of revolution, and there was no stable government with which an agreement could be made. King Edward was already a prisoner at Warwick's castle of Middleham in Wensleydale; his only army had been destroyed on 26 July at Edgecot near Banbury by an invasion of northcountrymen sponsored by Warwick, and Warwick and Clarence had beheaded the two most unpopular Woodvilles, the Queen's father Earl Rivers and her brother Sir John, at Coventry on 12 August. The Kingmaker had a new three-tier plan for fulfilling his life's ambition, to create a puppet king who would be his admiring friend and beg him to rule England. Preferably, he would humble

[1] 1469 by modern chronology, though Caxton whose year in Flanders ended at Easter says '1468'.

Edward into grateful and contrite obedience; if not, he would depose Edward, crown Clarence (whom he had married in defiance of Edward's veto to his daughter Isabel Neville at Calais on 11 July), and thereafter dominate Clarence as he had failed to dominate Edward; and if even Clarence should fail to see the light, he would restore Henry VI, the most submissive of all possible kings. The consequence of success for any of Warwick's alternatives would be an alliance with France and war against Burgundy, a gloomy prospect for the English merchants in Flanders.

Oddly enough, the merchants had just enjoyed their two most prosperous years since the beginning of the century. It might be expected that the repeated failure of the cloth negotiations would have depressed their trade, but the customs accounts tell a different story. In the two years since Michaelmas 1467 denizen cloth exports from London had suddenly doubled and tripled, soaring from the previous norm of about 8000 cloths to 15,052 in 1467-8 and 24,260 in 1468-9. This phenomenal rise was evidently due to the Anglo-Burgundian 'intercourse' treaty of February 1468,[1] and shows that Charles must have tacitly lifted the practical application of the cloth ban, or allowed the English merchants to evade it, although the treaty itself explicitly reserved the ban for future discussion. In fact, as time was to show, Charles was determined to keep the embargo as a bargaining lever, knowing that he could exert pressure whenever he pleased by threatening to re-enforce it. So the mixed outcome of Caxton's decade of diplomacy was an unprecedented practical success, offset by a metaphysical political hitch that could only be mended by the great ones who made it. He and his fellow-merchants had every reason to congratulate themselves, and Caxton perhaps now began to feel that his task was done.

Warwick's schemes were going awry. He had to set Edward free in October 1469, finding that the country could not be governed without him. Edward remained dismayingly uncowed, and the nation showed little sign of wanting to be rid of him, still less of preferring Clarence for king. Edward, Warwick and Clarence staged an effusive false reconciliation in December. Anthony Woodville came out of hiding, took his murdered father's title of Earl Rivers, and rejoined Edward's court. One of Edward's first acts after release was to take the oath of Charles's Order of the Golden Fleece, and to send Russell to Charles with the Garter. Caxton was probably present when Russell invested Charles with this gratifying but sometimes mis-understood emblem at Ghent on 4 February 1470;[2] if so, he doubtless

---

[1] But it was also intensified by the reduction of Hanse competition after the Verdict of July 1468.

[2] A Burgundian observer (as Vaughan notes) guessing wildly described it as 'the garter of the companionship of the ladies of England'. Charles had elected Edward to the Golden Fleece on 14 May 1468, just before the wedding, and formally announced the fact by a special embassy in April 1469, whereupon Edward awarded Charles the Garter on 13 May 1469. His choice of this moment for Russell's mission was an act of defiance towards Warwick, to show that he intended to keep close relations with Charles.

duly appreciated (for he was a competent Latinist) Russell's more than customarily tedious Latin speech on the occasion, comparing the Garter with King Arthur's Round Table, which Caxton himself was to print some years later.

In March Edward put down a rebellion in Lincolnshire which Warwick and Clarence had instigated, and proclaimed the guilty pair as traitors.[1] They narrowly escaped to France, where Warwick swore eternal fidelity to his mortal enemy Margaret of Anjou, and promised to restore Henry VI, with Edward Prince of Wales as regent; the Prince would marry his daughter Anne Neville, with succession to the throne passing to their offspring, if any; and Clarence, much to Clarence's disgust, now stood a poor fourth. In September they invaded England, and Edward fled to Holland, accompanied by the ever loyal but mutually antagonistic Gloucester, Rivers, and Hastings. He was driven upon the sandbanks near Alkmaar by Hanse warships and rescued by his old friend Louis de Gruthuyse, who gave him hospitality at the Hague. Warwick took Henry VI from the Tower, where he had languished ever since his capture by William Cantelow, mercer, in 1465, 'not so cleanly kept as should beseem such a prince', and made him King again. Edward's Woodville Queen took sanctuary at Westminster Abbey, where she gave birth to her first son on 2 November 1470; so now there were two Edwards Princes of Wales. For a moment Warwick was almost popular in a divided nation. The old nobility and other stay-at-home Lancastrians were his, and the squires and peasants from his vast domains in the north, the formerly Yorkist men of Kent whom he nurtured with Channel piracy as governor of the Cinque Ports, the masters and craftsmen of the lesser, non-mercantile, mercer-hating guilds, the London rabble whom he fed with beef from his kitchens and allowed to riot against foreigners. Edward's illogical subjects blamed the King for 'one battle after another, great loss of goods, and hurting merchandise', that is, for the civil wars, high taxation, and restrictionism which they themselves had forced upon him. Warwick, rather unexpectedly in view of his previous form, beheaded hardly anybody but the inhumane humanist John Tiptoft, Earl of Worcester, who had chopped off so many noble heads for Edward that he was called Butcher of England, and was missed by few except, curiously enough, Caxton himself.

Edward's situation seemed so hopeless, and his presence in Burgundian dominions so embarrassing, that for three months Charles would have nothing to do with him, except to send a small monthly pension for his upkeep. At last Charles awoke to the urgency of his own danger, finding that Louis had revived his old scheme for a simultaneous attack on Burgundy by France and England, to be followed by dismemberment; but this time Holland and Zeeland

---

[1] Edward's lightning victory at Empingham in Rutlandshire on 12 March 1470 was nicknamed Losecoat Field, for some of the enemy wore Clarence's livery, and threw it away as they fled.

would be given to Warwick, not Clarence. Now Charles suddenly remembered his dear brother-in-law Edward, sent for him, conferred with him on 2 January 1471 at Aire in Artois, granted him £20,000 for the voyage home, and even persuaded the angry Hanse to provide ships. Edward also saw Margaret, who pleaded with him to make a deal with her favourite brother Clarence. Then he stayed for nearly a month in Louis de Gruthuyse's mansion at Bruges, where visiting English staplers from Calais (they were always the ones with the ready cash) lent him money, while Rivers hired more ships. It is likely enough that Governor Caxton met his King for the first time on this occasion, and that he made himself useful in the quest for shipping, as he was to do again four years later for Edward's expedition against France in 1475. In February Charles parted at last with his star Lancastrian refugees, Somerset and Exeter, and allowed them to return to England—ostensibly out of deference to Edward's presence, and on the pretext that they would counterbalance their enemy Warwick, but really to hedge his bets by letting them fight Edward, as they did with the utmost vigour. When Edward sailed for England on 11 March 1471 it must have seemed that he, not Somerset and Exeter, was bound for death or captivity.

6

THE EXILE

CAXTON had now entered the most momentous and mysterious period of his life. Between the autumn of 1470 and the close of 1472 he ceased to be Governor of the English merchants and became a practising printer. It has generally been argued that he resigned his great post voluntarily and for personal reasons, to enter the service of Margaret of York, or to continue his literary work, or to learn printing, or to retire at the approach of old age, or to marry. But if we try to date the end of his governorship more precisely, and thereby to understand this event in its historical context, it becomes clear that the break in Caxton's career was forced upon him, and that his subsequent activities should be interpreted not as planned motives, but as unplanned consequences.

Caxton is last named as Governor ('master of the English nation') on a convivial but rather anticlimactic occasion in 1470 (unfortunately no month-date is revealed), when the town of Middelburg regaled him with the modest quantity of two quarts of Rhine wine and one of 'Gascon' (or Bordeaux). His successor was none other than John Pickering, the Yorkist mercer who had briefly replaced Overey in 1460-1, and served with Caxton on the abortive merchants-only delegation of September 1468. Pickering was already Governor on 5 March 1472, when he appears under that title in a list of trade envoys, but presumably his actual appointment had occurred some time earlier. The interval can be narrowed down still further. As we shall see, Caxton had already become free towards March 1471 to reside at the court of Margaret of York, and to resume the translation which he had laid aside for two years before; and in June 1471 he began eighteen months of residence at Cologne. Such a return to disponibility is hardly compatible with continuance of his office as Governor, and the dates strongly suggest that the end had come within, and as a consequence of, the critical six months of the Lancastrian restoration between October 1470 and March 1471.

During that topsyturvy interregnum the position of the English merchants in Burgundian dominions was dire indeed. They and their Governor, like the London Court of Adventurers to whom they owed obedience, were now suddenly subjects of Henry VI restored and Warwick the Kingmaker. In London the anti-alien lesser guilds and craftsmen for the home market had always favoured the economic nationalism of Lancaster or Warwick, and now triumphed, while the Yorkist oligarchy of mercers and Adventurers lay low. Even in Flanders a similar opposition faction of non-mercers and Adventurers from provincial centres existed; these had already enjoyed a brief ascendancy under the fishmonger Overey in 1456-9, and perhaps now appealed to the home government, as they later did against Pickering in 1478.[1] Meanwhile shipping to and from the forbidden land of Burgundy was at a standstill. The future prospect was even blacker, for Warwick was committed by his treaty with Louis to invasion of Burgundy, annihilation of Charles, abolition of Netherlands markets, and diversion of trade to France.[2] Caxton's quandary now resembled in reverse Overey's at the time of the Yorkist takeover ten years before. Perhaps he was dismissed, or even replaced, whether through an opposition coup at Bruges, as was Overey by Pickering in 1460, or by an intervention of the home government, as was Pickering when Overey made his brief comeback in 1462, or by order of the London Court of Adventurers, as when Caxton himself had ousted Overey. Or perhaps, finding his position untenable or intolerable, he resigned, or stood aside and let his deputy take over, as did the Yorkist Mayor of London in that same February.[3] But however this may be, it seems certain that Caxton's governorship ended at this time, as a direct consequence of the Lancastrian restoration, and without his wish or intention.

To Caxton this sudden loss of his post must have seemed disastrous. For nine years he had served his fellow merchants as an administrator and protector, and the State as a diplomat, in the most prominent, powerful and lucrative office available to a private Englishman abroad. He could not foresee that his misfortune would turn out to be a blessing in disguise, and lead to his second and (in the eyes of posterity) far more illustrious career as a printer; indeed, it would be anachronistic to suppose that Caxton had ever considered printing when still Governor, or even in the first months after his fall. But perhaps, as sometimes happens, the psychological upheaval caused by a broken career made him unconsciously ready for a change of vocation. Meanwhile he could do no better than cultivate his acquaintance with the Duchess Margaret, and solace his involuntary but perhaps not quite unwelcome leisure by resuming his translation.

Caxton's association with Margaret had very probably begun at

[1] Pickering was rebuked in 1478 by Edward himself, in response to a petition from the Adventurers of York, Hull, Scarborough and other northern ports, who complained of his demanding exorbitant fees from non-Londoners.

[2] The customs records show that in 1469-70, the first year of Warwick's supremacy, cloth exports from London had already dropped to less than half the 1468-9 peak. A gap in the records during the Lancastrian interregnum probably reflects an almost complete break in trade, while a spectacular rise immediately after Edward's restoration shows a crash programme to ship the backlog so created.

[3] Mayor John Stockton, a mercer, faced with defending London against Edward, fell diplomatically ill in February 1471, and relinquished his post to Sir Thomas Cook, a draper and former Yorkist who had fallen foul of the Woodvilles, turned Lancastrian, and stood trial for treason in 1468 after lending money to Margaret of Anjou. When Edward landed Stockton suddenly got well again and was rewarded with a knighthood.

the time of her marriage to Charles in 1468, and continued ever since. He had served on diplomatic missions throughout the 1460s with the same high Yorkist dignitaries, including her brother-in-law Rivers, Wenlock, Vaughan, Hatcliff, Russell, who had negotiated her match and attended her wedding, and he was head of the English colony who welcomed her to Bruges. No resident Englishman was more fitted by experience, position, contacts and loyalties to act as her political and commercial adviser in her strange new home of Burgundy, where he had lived for more than twenty years. Margaret, like all Yorkist royalties, dabbled ardently in the cloth trade, for which her brother Edward granted her licences to export duty free, and Caxton as Governor of the English merchants may well have served as her agent. Then again, Margaret was a patroness of authors and scribes, a devoted reader of chivalric romance and pious literature, like Caxton himself, and no doubt delighted to welcome this uniquely literate merchant to her court. However, it is clear from Caxton's own account that he was already being summoned to Margaret's presence to discuss 'diverse matters', and receiving her bounties including an annual fee, before he resumed his translation at her request, or even told her of its existence; so their relationship was not merely one of author and patron, as Blake argues. Nor, on the other hand, can he have been a resident fulltime employee in her court, as some have supposed, for his work and movements as governor-diplomat would have precluded this. The actual situation probably was that, while still Governor, perhaps ever since 1468, he took a retaining fee as her adviser and business agent, and performed these tasks by occasional visits, but mostly by letter or messenger.

So, when the Duchess sent for the ex-governor towards March 1471 'to speak with her good grace of diverse matters', as Caxton tells us in his prologue to the *Recuyell*, they no doubt first discussed the parlous state of politics and trade, Warwick's iniquities and Edward's chances, and then turned to literature. This was Caxton's opportunity. 'I let her highness have knowledge of the foresaid beginning of this work, which anon [i.e. immediately] commanded me to show the said five or six quires to her said grace, and when she had seen them she anon found a fault in mine English, which she commanded me to amend, and moreover commanded me straitly to continue and make an end of the residue then not translated; whose dreadful commandment I durst in no wise disobey, because I am a servant unto her said grace and receive of her yearly fee and other many good and great benefactions, and also hope many more to receive of her highness, but forthwith went and laboured in the said translation after my simple and poor cunning.'

When and where did this famous interview occur? Caxton implies a date towards March 1471, when he says he began his translation at Bruges on 1 March 1469, abandoned it after writing 'a five or six quires', and 'in two year after laboured no more in this work', until Margaret ordered him to continue. This date follows so closely upon the period which we have deduced on political and documentary grounds for the end of his governorship, that the two tend to confirm one another. Further evidence not only supports this chronology, but reveals the place. In his epilogue to the second of the three books of the *Recuyell* Caxton mentions that his work up to this point was 'begun in Bruges, and continued in Ghent, and finished in Cologne'. Why Ghent? Caxton's business had made him familiar with this centre of the Flemish weavers ever since the 1450s, and he may well have been present there a year before, when Russell invested Charles with the Garter on 4 February 1470. But it is more significant in this context that Margaret herself often lived at Ghent in her palace of Ten Walle. Thanks to the researches of the Belgian scholar Van der Linden the Duchess's movements and whereabouts are known for every day of her married life. As it turns out, after seeing her brother Edward at Aire in January 1471, and a brief sojourn at Lille, where her brother Gloucester called on her, she moved to Ghent on 16 February, and stayed there till 25 June. So it was at Ghent that Caxton's interview took place; and our new data also endorse the date we have assigned to it, and explain why the *Recuyell*, in obedience to Margaret's command, was 'continued in Ghent'. The Duchess had evidently given her fee-earning adviser and ex-governor refuge in her court. Probably his visit began (if our conjecture that he helped Rivers to find shipping for Edward at Bruges is justified) soon after Edward sailed for England from Flushing on 11 March. He stayed on, industriously translating,[1] until a few days before Margaret's departure, and then moved to Cologne, where he first appears, as we shall see, on 20 June 1471.

Meanwhile Edward had made his Napoleonic return from exile, and recovered his kingdom in the hardfought, doomladen, crushing victories of Barnet on 14 April and Tewkesbury on 4 May. False, fleeting, perjured Clarence joined him with an army raised for Warwick. Warwick, slain in flight from Barnet, was displayed a naked corpse in old St. Paul's. Edward, Prince of Wales,[2] was captured and done to death at Tewkesbury, by Gloucester himself as some said, and crying to Clarence for mercy. Charles's favourite Lancastrian Somerset was dragged from sanctuary in Tewkesbury Abbey and beheaded among the victims of Gloucester's bloodbath. He had fought in the van with desperate courage against Edward and Gloucester in person, till his men broke and were slaughtered in the field still called Bloody Meadow. Wenlock, Warwick's man and Caxton's former diplomatic chief, had failed to support him, and Somerset, reviling him for 'standing still', had called him traitor and

[1] An approximate calculation suggests that he translated this lengthy book (about 210,000 words comprising 350 leaves or 700 pages in its printed form) at the rate of 50 leaves or 30,000 words a month, and that half of his task was done at Ghent.

[2] Aged seventeen. According to the Milanese ambassador to France in 1467 the Prince then already talked 'of nothing but cutting off heads and making war, as if he was the god of battle or the peaceful occupant of the throne'.

with his axe 'struck the brains out of his head'. Rivers stayed in London with Mayor Stockton (who suddenly returned to health) to beat off a rising of Warwick-supporting Kentishmen, not unlike Cade's in 1450, led by Thomas, Bastard of Fauconberg,[1] and favoured by Canterbury and Cinque Ports merchants. Edward reached London on 21 May (when the Bastard had just fled, after bombarding the city and firing London Bridge), and immediately sent Gloucester 'to make a bloody supper in the Tower'. Next day it was announced that Henry VI had suddenly died, not of the dent in the back of his skull which was said to be found when he was exhumed in 1911, but of 'pure displeasure and melancholy'. 'My cousin York, I know that in your hands my life will not be in danger,' poor Henry had said on re-entering the Tower, and Edward had promised he had nothing to fear. So Gloucester, aged eighteen, had now learned evil lessons. Many thought the murdered King a martyred saint, and as late as 4 November 1480 the Court of Adventurers had to forbid members of its fellowship to make the politically undesirable pilgrimage to Henry's tomb at Chertsey. Fauconberg (after first being pardoned and then breaking parole) was beheaded by Gloucester on 22 September 1471, and his head set on London Bridge 'looking into Kentward'. Nicholas Faunt, grocer, mayor of Canterbury, was hanged, drawn and quartered to encourage the others, and a few more Kentishmen were executed or imprisoned; but mostly Edward did a roaring trade in fines or pardons, which both the guilty and the apprehensive innocent hastened to accept, £100 for a rich man, seven shillings for the poorest, 'so the King had out of Kent much good and little love'. Thomas Caxton the younger of Tenterden took out a pardon that autumn, like Hugh Caxton of Sandwich before him in Jack Cade's year of 1450, and our William Caxton, as we shall see, followed suit in the spring of 1472. Charles's other protégé Exeter, wounded and stripped for dead at Barnet, was rescued to sanctuary at Westminster Abbey, where even Warwick had left Edward's Queen and baby Prince undisturbed; but Edward had him out and put in the Tower.[2] Margaret of Anjou surrendered, and she too went to the Tower, harmless for ever.[3]

Duke Charles heard the not altogether welcome news of Barnet from Duchess Margaret within four days of the battle, together with

---

[1] The Bastard, being an illegitimate son of William Neville Lord Fauconberg, Earl of Kent, a brother of Edward's mother Cecily Neville Duchess of York, was like Warwick himself a distant cousin of Edward. Many Kentishmen were still Warwick Yorkists rather than Edwardian Yorkists, and one of the earliest symptoms of Warwick's bid for power was a local raid on a Rivers estate in Kent (the Mote, near Maidstone) on 1 January 1468. However, a strong force of Kentishmen fought on Edward's side at Barnet, where their leader Sir Thomas Bourchier was slain.

[2] Edward let him out to join his French expedition in 1475. Exeter mysteriously fell overboard on the voyage home, and it was thought that Edward had arranged for him to be pushed.

[3] Edward ransomed her to Louis in 1475. Louis took her property and gave her a tiny pension. She died in 1482.

a message that Edward was far from satisfied with the help he had received. After Tewkesbury Edward wrote more graciously, proposing reciprocal renewal of their treaty of trade intercourse, which the Lancastrian restoration had rescinded, and sent thank-you letters to Louis de Gruthuyse and the city of Bruges for their hospitality. Charles proclaimed public rejoicings, and Caxton no doubt took a day's holiday from the *Recuyell* on 12 June to attend the huge bonfire at Ghent with which Duchess Margaret celebrated her brother's victory.

But the tremendous events of that spring had come too late to save Caxton's former career or to divert him from his new one. When the Duchess left Ghent for Brussels on 25 June 1471 Caxton was already in Cologne, where entries in the aliens' register show his presence during the next eighteen months, until the end of 1472. The date of the first of these, 17 July 1471, has always been misinterpreted as showing the time of his actual arrival. In fact the entry grants him permission to remain in Cologne for a further month 'with three days in hand' (*'cum resignatione iij dierum'*), so indicating that he had then already been there for a calendar month all but three days, that is, since 20 June. Evidently he had arrived, as was usual when entering a foreign city, with a month's safeconduct or letter of protection (the ancestor of the modern passport), which in this case would have been issued by the Duchess, or by the City or Bishop of Ghent; and he chose to obtain his first residence permit proper in good time, three days before the safeconduct expired, on 17 July instead of 20 July, but was allowed to keep his 'three days in hand'. Similarly when he renewed his permit on 9 August, eight days early, he was given until Christmas 1471, 'with eight days in hand'. His two next renewals, one on 11 December 1471 until 24 June 1472, and the last on 19 June 1472 'for half a year', were each made a few days early; but on both occasions, instead of complicating matters by cumulative carrying forward, he received only the eight days still left in hand from the August entry (*'ut supra cum resignatione 8 dierum'*).

Duchess Margaret's removal from Ghent on 25 June 1471 provides an adequate explanation for both the timing and the cause of Caxton's departure five or six days earlier. Evidently at that moment, after about four months as a resident in her court, he was obliged to seek a new home. But why Cologne? So far as is known, Caxton had never been there before, but he would certainly have made many contacts during his three decades abroad with the Hansards of Cologne, particularly at the time of his exodus to Utrecht in 1464-7 and the Hanse negotiations of 1469. In 1473-5 we shall find Caxton engaged in further diplomacy with the Hanse, with special reference to that of Cologne. Was there any such task in the summer of 1471, for which his services would be required? Yes, to be sure! Edward was then about to resume his typically devious efforts to reach agreement with the Hanseatic League, to whom he had promised 'great privileges' when they ferried his army from Flushing that March. One of their chief demands was the expulsion from London of the rebellious Hanse of Cologne, to which Edward could not agree, for Cologne was essential both as a bargaining lever in the future negotiations, and as the only remaining Hanse outlet for English cloth.[1] One of the King's first actions, after dealing with the men of Kent in May, was to confirm yet again, in June and September, his 1469 grant of exclusive privileges to Cologne.[2] So, when we find Caxton repairing to Cologne at this very time, it seems more than likely that he was instructed to present the King's patent of renewal to the Hansards of Cologne, to reassure them of Edward's good intentions, and to continue to act as a link man in preserving the Cologne connection.

Such a mission is not incompatible with the co-existence of another motive. Caxton's long sojourn in Cologne may also have been in the nature of 'a voluntary and protective exile'. On 8 March 1472 he saw fit to take out one of Edward's expensive but all-inclusive pardons, covering 'treasons, murders, rapes of women, insurrections, rebellions, conspiracies, riots, felonies, burglaries and other malefactions', in the name of 'William Caxton of our city of London, mercer'. No doubt our hero was personally blameless of any of these horrors, for the document embodied a bureaucratic formula applied to guilty and innocent alike. Perhaps it was no coincidence that Thomas Caxton of Tenterden (whether or not he was our Caxton's brother) had obtained a similar pardon a few months before, and our Caxton may have been involved in some indirect, long-distance way with the Kentish troubles of 1471. He may, for example, have owned property in Canterbury or elsewhere in Kent, and had reason to fear confiscation for the misdeeds of his fellow proprietors. Caxton can hardly have been a willing partisan of the usurper Warwick, still less a crypto-Lancastrian, for their policy of abolishing Burgundian markets and diverting trade to France meant ruin for himself and the Adventurers whom he had protected as Governor and diplomat. But he had served with Warwick men, including the slain Kingmaker himself and Wenlock, and others still alive such as Whetehill, in the years when Warwick was still allowed by Edward to pursue his anti-Burgundian plans. Perhaps Caxton had remained involved with these, or sought their favour as a safety measure during the interregnum. Perhaps the troubles which ended his governorship had left him with enemies in London or Bruges, against whose threats he defended himself with exile and royal pardon. Two circumstances suggest that he intended his pardon as a safeguard for an eventual return to Bruges. He obtained it at the very moment when, on 5 March 1472, his former Yorkist colleagues Hastings, Hatcliff,

---

[1] It is a measure of Cologne's importance that during the Hanse war and ban from 1470 to 1475 the Hanse cloth exports from England still remained (as customs records show) at between a third and half of their former level.

[2] The 1469 grant had already been renewed at Easter 1470 and again, surprisingly enough, early in 1471 during Henry's restoration.

Russell, and the new governor Pickering were commissioned to attend the first postwar diet at Bruges on peace, trade, and the Hanse problem, for which his attendance might be required. Still more significantly, by this time he had already begun to learn printing at Cologne, with a view to setting up his own press at Bruges.

7

LEARNING TO PRINT AT COLOGNE

CAXTON'S business at Cologne, whatever it was, proved less absorbing than the new fine delight of literature. He pursued his translation at the same rate as in Ghent, a thousand words day in day out, which is good going for any writer, especially for one in the first year of retirement. Towards mid-July he completed the second of the three books of the *Recuyell* with an interim epilogue of his own. He had intended to omit the third book (which dealt with the final destruction of Troy) as superfluous, 'for as much as that worshipful and religious man Dan John Lydgate monk of Bury did translate it but late,[1] after whose work I fear to take upon me that am not worthy to bear his penner and ink horn after him to meddle me in that work'. Instead he pressed on, pleading in selfexcuse that Margaret had ordered the whole, that some readers like prose better than rhyme, 'and also because that I now have good leisure being in Cologne and have none other thing to do at this time.' But in fact he could not help himself, and this moment when he had meant to halt and found he could not marks the point of no return in his vocation as translator. Even so, he was not yet free from regret for his lost eminence, and resentment for the troubles that caused his semi-exile. There is a note of rueful irony in his 'good leisure being at Cologne'. Other words in the same epilogue are still more revealing. He has written, he laments, in the year 1471, 'in the time of the troublous world and of the great divisions being and reigning as well in the realms of England and France[2] as in all other places universally through the world'. For a formerly staunch Yorkist, who might be expected to express reprobation of the traitor Warwick, rejection of the usurper Henry, joy for Edward's triumph, these are strangely neutralist sentiments. But then, he was writing in a foreign city for the Clarence-loving consort of an Edward-hating Duke, at a time of personal humiliation and disillusion.

Caxton finished his final stint of 60,000 words on 19 September 1471, and no doubt showed the Duchess his book soon after.[3] Raoul

---

[1] Lydgate's *Troy Book*, an English narrative poem written about 1420, was quite unrelated to the French prose *Recueil*; however, both works were based on earlier Latin sources and were therefore regarded by Caxton as translations.

[2] Louis's problems in the summer of 1471 seemed not unlike Edward's in the spring, for he was threatened by a hostile league comprising his own brother Charles Duke of Guienne in Normandy, Francis II of Brittany, and the Count of Armagnac and Gaston de Foix in the south.

[3] 'Which book I have presented to my said redoubted lady,' he says in the final epilogue written at the time of printing about three years later, 'and she hath well accepted it and largely rewarded me.' The total 210,000 words

Lefèvre's *Recueil des histoires de Troie* was the latest Burgundian best-seller, for the first version in two books only had appeared in 1464, while the version augmented with a third book and translated by Caxton was more recent still. The author, as Lefèvre says himself, was 'priest and chaplain' (but not, as Blades says, secretary) to Philip the Good, at whose command he wrote the book, just as Caxton translated it at Margaret's. Nothing more seems to be known of his life, but he must surely have been a son or nephew of the Jean Lefèvre who was Philip's chamberlain, councillor, ducal herald, Golden Fleece king-at-arms, and wrote a chronicle of his times. Raoul Lefèvre had also produced a successful *Histoire de Jason* for Philip, a romance of obvious Burgundian topicality on the classical legend of the Golden Fleece, which had provided Philip with the theme of his order of knighthood long before in 1430. Caxton knew this earlier work equally well, and was to print it a few years later both in the French and in his own English translation. But the tale of Troy, strangely enough, ranked almost as high as the Golden Fleece in Burgundian state mythology. By one of those delightful false etymologies of which the middle ages were so fond, the ancient Trojans or *Teucri* (as Virgil called them) were now inextricably equated with the modern Turks or *Turci*, whose assaults on the Greek Empire were regarded as a retaliation for the Trojan War! Everyone took this preposterous notion in deadly earnest. In 1444-5 Philip had sent a Burgundian fleet to the eastern Mediterranean with instructions to visit the sites of Hector's Troy in Asia Minor near the entrance to the Hellespont and Jason's Colchis at the furthest southeastern corner of the Black Sea, land of the Golden Fleece, as they duly did; and the Sultan himself responded by a menacing letter to Philip, signing himself 'King of Troy and true heir of Hector of Troy'. 'The Turks have indeed avenged the taking of Troy,' remarked the Catalan traveller Tafur when Constantinople fell in 1453. Until his dying day Philip threatened to go on crusade against the Turks as soon as he had time, and made his entire court swear to come too.[1] Englishmen might disagree, but only because they were equally convinced that London was founded under the name of Troynovant or New Troy[2] by Brutus, the great-grandson of Aeneas the Trojan.

---

would presumably have to be copied first by a professional scribe. Duff, followed by Crotch, asserts without qualification that Caxton presented his book 'at the close of 1471'. In point of fact Caxton does not specify the time, and the context implies that he refers here to the printed book in 1474-5. Still, it is reasonable to assume that he did show Margaret his work in manuscript soon after completion. The Duchess was at Brussels in June to October, at Arques in November 1471 to January 1472, and thereafter at Ghent.

[1] On 17 February 1454, at the famous Vow of the Pheasant (because the pheasant was the Phasian bird from Jason's Colchis), with festivities including the story of Jason in dumb show.

[2] Another false etymology, from the tribal name of the Trinobantes who inhabited Essex in Julius Caesar's time. Brutus, it was felt, was a highly suitable name for the first Briton.

The demand for Caxton's English *Recuyell*, irrespective of whether Trojans were ancient Britons or modern Turks, proved greater than could be met by manuscript alone. 'And for as much as in the writing of the same my pen is worn, mine hand weary and not steadfast, mine 'eyes dimmed with overmuch looking on the white paper' (as he says in words that remind us of both Mallarmé and Mr Pepys) 'and my courage not so prone and ready to labour as it hath been, and that age creepeth on me daily and feebleth all the body, and also because I have promised to diverse gentlemen and to my friends to address to them as hastily as I might this said book, therefore I have practiced and learned at my great charge and dispense to ordain this said book in print after the manner and form as ye may here see.'

Caxton's own account of the most crucial event in his life, his decision to learn printing, ought not to be lightly rejected. It is logically and psychologically convincing, and fits with all the facts against which it can be checked, including our new evidence concerning the time and circumstances of his loss of office and his sojourn at Ghent. These events must surely have preceded his resolve to learn printing, and we may accept his statement that he reached this decision at Cologne in the autumn of 1471, after completing the *Recuyell* on 19 September, and after the gratifying discovery that 'diverse gentlemen'[1] and his friends would be eager to acquire their own copies.

The new craft of western printing, then only two decades old, had been invented at Strassburg in the late 1430s and perfected at Mainz in the early 1450s by the Mainz patrician and goldsmith Johann Gutenberg (*c.* 1398-1468). Gutenberg's invention consisted of course not of the mere idea of impression from inked letter-forms, which was obvious to all men, but of the technical means of doing so with efficiency. The fundamental and unique new features of his discovery included an adjustable handmould, with punch-stamped matrices, for precision casting of type in large quantities; a type-metal alloy (probably lead with tin and antimony) with low melting-point and quick, undistorted solidification; a hand press adapted from those used by, among others, papermakers and bookbinders; and an oilbased printing ink. Gutenberg was a towering genius; his 42-line Latin Bible produced about 1453-5 remains perhaps the most beautiful and technically perfect book ever printed, and printing by hand-press neither required nor received any radical improvements for three centuries to come.

Printing spread only slowly during the first generation, first along the Rhine to Strassburg (1460), Cologne (1465), Basel (about 1468), then over the Alps to Subiaco (1465), Rome (1467), Venice (1469), and in 1470 to Paris. Proliferation accelerated in geometric progression as new presses were founded in new towns by partners who severed company, employees who left their masters, or feepaying pupils who became proficient. Some masters supplied

---

[1] Misquoted by Crotch as 'many lords'.

newcomers with type ready made or designed to order. Entre-
preneurs in the existing manuscript trade or merchant investors
often provided finance, markets, and managerial organisation. All
these typical processes are already found in Gutenberg's career.[1]
When Caxton reached Cologne in 1471 printers were at work in a
score of towns in Germany and Italy, and the expansion of printing
was about to become explosive for the next two decades.

No doubt Caxton had been well aware of the existence of the new
art for several years. He would have heard from his Cologne Hansard
contacts of the first press in their city; the anonymous and mysterious
Dutch Prototypographer, the first printer in Holland, was already at
work in or before 1466, possibly at Utrecht at the time of Caxton's
exodus; and as we have seen, printed books were already on sale in
Bruges as early as 1467. But Caxton can hardly have thought
seriously of learning to print while he was still Governor, or indeed
at any time before his resumption of the *Recuyell* in Margaret's
palace at Ghent; and the idea, though perhaps present in his mind,
can hardly have predetermined his move to Cologne, which is
adequately accounted for by other motives. The moment of decision,
the conscious acceptance of his vocation, the reasoned estimate of its
practicability, surely came at Cologne itself, in the last months of
1471, when he had finished his translation, saw printers at work, met
their satisfied customers, and spoke to the master-printer who was
willing to take him as a paying pupil.

Three separate printers were active in Cologne at that time. The
first was Ulrich Zel, a cleric from the diocese of Mainz, who began in
1465, the second Arnold ther Hoernen, who began in 1470, and the
anonymous third commenced soon after ther Hoernen (or perhaps
even before, in 1469). Which of these was Caxton's teacher? The
answer is given by none other than Caxton's foreman and successor
Wynkyn de Worde, who ought to know. Towards 1495, about four
years after Caxton's death, de Worde produced the first edition in
English of the thirteenth-century encyclopaedia of Bartholomaeus
Anglicus, *De proprietatibus rerum*, with a verse epilogue which includes
the celebrated lines:

> *And also of your charity call to remembrance*
> *The soul of William Caxton first printer of this book*
> *In Latin tongue at Cologne himself to advance*
> *That every well disposed man may thereon look.*

Now the first Latin edition of *Bartholomaeus* was in fact printed at
Cologne in 1472, and there can be no doubt that Caxton learned his
craft by taking part in the production of this book. So his master

[1] Gutenberg contracted to teach printing to Andreas Dritzehn at Strassburg
in 1438; he left his Mainz partners, the capitalist Fust and the type-designing
Schoeffer, in 1455 to print independently; his former employees Berthold
Ruppel and Heinrich Kefer founded their own presses; he supplied type and
tuition to Albrecht Pfister at Bamberg in 1459 and to Heinrich Bechtermünze at
Eltvil towards 1467.

1 Death among the Printers. The earliest known illustration
of a printing press, from Danse Macabre, Lyons, 1499.

was the anonymous printer of the *Bartholomaeus*, who is generally known as Printer of *Flores sancti Augustini*, from a book produced in the following year, 1473. The type used by this printer is found at Cologne in various slightly different states, the first (or rather, second?) with Arnold ther Hoernen in 1470, and others in various groups of books which were formerly assigned to separate presses, the Printer of *Dictys* (1471—or rather, 1469?), Printer of *Dares* (1472), and so on. But more recent study of the type seems to show that these distinctions are not valid for differentiation of presses, for the supposed separate states of the type tend to overlap from group to group, and even to occur in one and the same book.[1] This suggests that all these books may have been produced in a single large establishment, using several presses with slightly variant cases of type supplied by one and the same typefounder.

Modern typographical scholars have succeeded in identifying this typefounder by name, using the method of tracking his later use of the type, like footprints in sand, until at last he reveals himself as a known person. The type reappears, in a state almost identical with the Cologne *Flores sancti Augustini*, in the *Lis Belial* printed at Louvain soon after 7 August 1474 by the wellknown printer and typefounder Johann Veldener, a native of the diocese of Würzburg in Bavaria. Veldener became one of the most distinguished and prolific makers of type in the Low Countries, and Caxton himself (as we shall see) was to be one of his chief customers. So it begins to seem very likely that Veldener was Caxton's teacher at Cologne in 1472. But we cannot assume this unless we can feel sure that Veldener was the actual working master of the *Flores* press, and not merely its supplier of type.

An alternative candidate as master of the *Flores* press has been suggested in the person of Johann Schilling, who printed at Basel in 1473-6 using a state of the type indistinguishable from that found in the *Flores*; so that Schilling, typographically speaking, might seem to have as good a claim as Veldener to be regarded as Printer of the *Flores*. Schilling had apparently worked at Cologne before moving to Basel, for the presence in Basel University Library of several books from the *Dares* group no doubt means that these were originally acquired from Schilling himself. Presumably Schilling received a fount of the type, and a number of the books he had helped to print, as his share of the assets when he left the Cologne press.[2] But

[1] This conclusion was reached by my colleague L. A. Sheppard in an unpublished article written towards 1954, but seems already implicit in the detailed analyses of the type in BMC I. The same article shows the strong probability that the anonymous press commenced *before* Ther Hoernen's, i.e. in about 1469.

[2] One school of thought even argued from this evidence that the *Dares* group was actually printed at Basel; but Allan Stevenson has proved their Cologne origin from their paper, which came from Épinal and Lorraine, and would be transported downstream to Cologne via the Moselle, but would not be shipped upstream to Basel. Conversely Stevenson shows that the *Albertus Magnus* group assignable to Schilling at Basel is printed on paper made at Basel itself.

Schilling, although he was proud of his degree of M.A. (1465) at the University of Erfurt, is a much less impressive figure than Veldener. He could not cast his own type, and when his *Flores* fount gradually wore out was obliged to mix it with letters from a local Basel type. He fled from his Basel creditors in 1476, turned up again in France at Vienne (twenty miles south of Lyons) to print a few small tracts in 1478-9, commissioned a book from a Toulouse printer towards 1485, and was still wanted for debt at Basel in 1490. Veldener, on the other hand, belonged (as he already proclaimed in the 1474 *Lis Belial*) to the relatively small and select class of 'master printers'. A master printer (if we may define this ambiguous and obscure term from the manner in which it was generally used) claimed mastery in several senses: he was his own master, being an independent publisher-printer, not an employee; he possessed mastery of his craft, including the cutting and casting of type; and he was competent and willing to teach others as a master. The status of master printer in every sense with which Veldener emerges from anonymity after leaving Cologne, together with his position as Caxton's supplier of type, seems decisive for his identification as the master-mind of the *Flores* press and as Caxton's teacher. Schilling M.A. may well have been the scholarly partner in the Cologne press, but was no technical or teaching master.

The *Bartholomaeus* on which Caxton learned printing, a tall folio of 248 leaves and about 300,000 words, is the finest and largest book in the *Flores* group, and is almost unparalleled in Cologne printing at that date.[1] Even so, it is only one among the six expensive folios produced in 1472-3 by the *Flores* press, whereas the earlier *Dictys* and *Dares* groups consist only of cheap small quartos. Perhaps the finances of the press were rejuvenated by Caxton's learning 'at my great charge and expense'. But this sudden expansion into folios was itself a response to the competition of a newcomer to Cologne printing, Johann Koelhoff the Elder, who began in 1472 to produce bulky folios in a series of elegant types of the latest Venetian design. In fact the new specialist activity of the *Flores* group marked the death throes of the establishment. Towards the middle of 1473, evidently because its resources were overstrained by unsucessful competition with the formidable Koelhoff, the anonymous press ceased production and went into liquidation. Schilling moved to Basel, Veldener to Louvain, while a remnant of the type made a final appearance at Cologne in 1475 with a certain Goiswin Gops.[2]

Veldener's move to Louvain holds important consequences for the

[1] The only previous Cologne book of similar size and format is an Aquinas, *Summa theologiae*, produced by Zel about 1468.

[2] It is uncertain whether Gops was a minor member of the original staff who stayed on, or an outsider who acquired a remainder of the type stock after the liquidation. Like Schilling he had to resort to makeshift in replenishing his type, so he was not a typefounder. Arnold ther Hoernen, whose connection with the press was apparently confined to Veldener's supplying his type (and tuition also?), carried on until his death in 1482-3.

date and circumstances of Caxton's beginnings at Bruges. It has long
been supposed that Caxton began printing at Bruges in 1473 and
Veldener at Louvain in 1474, with no apparent relationship of cause
and effect. Now it will be shown from new evidence that Veldener in
fact founded his Louvain press in 1473, that Caxton did not commence
at Bruges before 1474, and that the two events are causally connected.

8

PRINTING AT BRUGES I: MARGARET AND CLARENCE

CAXTON'S last appearance in the aliens' register at Cologne on 19 June 1472 is conventionally interpreted as giving him permission to reside until the end of the year.[1] Whether or not this is correct, there is no good reason to doubt that he must have remained in Cologne until December 1472 or even a little later in order to see out the enormous *Bartholomaeus* and finish his tuition. Next it is generally assumed that he moved to Bruges a fully fledged printer, complete with type, trained his staff and set up his press, and finished the *Recuyell* edition early in 1474, or even before the end of 1473. But this natural-seeming hypothesis involves more than one serious difficulty, and needs to be reconsidered, with results that compel a revised view of the foundation and chronology of Caxton's Bruges press.

Caxton's next book in the English language after the *Recuyell* was the *Game of Chess*, which he finished translating on 31 March 1475,[2] and no doubt printed not long after. This, if we follow the old view that the *Recuyell* was completed early in 1474, if not sooner, leaves an unexplained gap of a year or more between the two books. In an attempt to fill this gap, it has been suggested that the three undated French books in this type may have been printed during the interval. But Allan Stevenson's study of the paper in the *Recuyell* and *Game of Chess* has proved that these two English books were in fact printed consecutively without an interval.[3] Consequently Caxton must have

[1] The entry reads: 'William Caxton from England, permit continued as above for six months [*ad medium annum*] with 8 days in hand'. *Ad medium annum* is rather strange Latinity, for it ought to mean 'until mid-year', i.e. till midsummer. Was the clerk merely confirming Caxton's previous permit, which ran till St. John's Day (24 June) 1472?—in which case Caxton presumably obtained a renewal which somehow failed to get registered. Or was he calculating from the commencement of Caxton's residence on 20 June 1471, in which case the next mid-year would indeed expire at the end of 1472?

[2] In his epilogue to the *Game of Chess* Caxton says '31 March 1474'; but as the Flanders year was reckoned from Easter, which fell on 10 April in 1474, the year must be 1475 by modern reckoning.

[3] Stevenson's results show that Caxton printed each of the three books of the *Recuyell* simultaneously on three presses, a fact already evident from the arrangement of the quiring into three sections. Each press began with a stock of the same paper (from Piedmont, with a grape watermark), and each shifted to a second paper (from the Vosges, with a bull's head watermark) when the first was exhausted. This second paper was continued in *Game of Chess*, which must hence have been printed immediately after the *Recuyell*. The grape paper was also used in a book (Thomas Cantipratensis, *Bonum universale*, c. 1473) belonging to the

started work on the *Recuyell* a year later than was thought, in late 1474 not 1473, and the establishment of his press after leaving Cologne must have occupied nearer two years than one. This delay is less surprising than it may seem. Caxton's training at Cologne doubtless gave him sufficient practical experience of printing processes and organisation to enable him to superintend his own office once it was in operation. But the founding of a new establishment required also the construction of presses and other equipment, and the training of a numerous staff of pressmen (two to each press), compositors, and other specialists. Even a master printer could hardly undertake all this in less than a year, and it is most unlikely that Caxton, as a novice gentleman printer, was ever competent to accomplish it at all.

The natural explanation of the difficulty would be to suppose that Veldener not only taught Caxton at Cologne but also set up Caxton's press at Bruges. This seemed impossible, so long as it was believed that Caxton began at Bruges in 1473, and that Veldener did not leave Cologne for Louvain until the summer of 1474, when he printed the *Lis Belial*. But in fact Veldener must surely have left Cologne when the anonymous press broke up, towards mid-1473; and he was already in Louvain on 30 July 1473, when he matriculated in the faculty of medicine at the University. Such a matriculation was a common practice with printers in university towns all over Europe, not necessarily with the intention of proceeding to a degree (a few did, like Schilling, but most, like Veldener, didn't), but in order to gain official status and privileges in the selling of books. Another discovery due to Allan Stevenson shows that Veldener began printing at Louvain sooner after his matriculation than was hitherto recognised. An undated edition of Boccaccio, *Genealogia deorum*, in a later version of the *Flores* type which must come immediately before the 1474 *Lis Belial*, turns out to be printed on paper made at Bar-le-Duc and Troyes which would normally be transported down the Meuse to Louvain, but would not find its way to Cologne; so Veldener must have printed this book at Louvain late in 1473 or early in 1474, as the first product of his new press. From Louvain it would be easy for Veldener to supply Caxton in 1473-4 with presses, equipment, and trained staff, and to visit Bruges, only 70 miles away, from time to time to supervise their installation. The relatively small output of Veldener's Louvain press (little over two books per year in 1473-7) shows that he would not lack time for this task. The setting-up of new presses for other men was apparently, like the supply of type, one of the financial mainstays of Veldener's enterprising but rather precarious career. There seems no reason to doubt that it was Veldener

Augustinus de Fide (Goiswin Gops?) group of the anonymous Cologne press, and Caxton perhaps ordered if from the same paper merchant. The bull's head paper occurs in Petrus de Crescentiis, *Opus ruralium commodorum*, printed by Veldener's rival Johannes de Westfalia at Louvain, 9 December 1474, which suits our new later date for the *Recuyell* very well.

who organised the press of the Brothers of the Common Life at Brussels in 1474, for which he certainly made the type, and it is probable that he did the same for others of the dozen or so new presses for which he supplied type during the next twelve years. Perhaps the prospect of Caxton's second fee was among Veldener's inducements (besides the attractions of a university town) for moving to the Low Countries after the dissolution of the Cologne press.

The provision of trained staff presented special problems. Perhaps Veldener was able to bring in jobless surplus craftsmen from the defunct Cologne press. But these would be Germans, whereas Caxton was about to print in English, and probably already foresaw that he would soon need to print also in French. His compositors, at least, would have to be familiar with these languages. Perhaps he arranged for Veldener to train a few likely lads from the English community at Bruges. He would also need an expert foreman, or supervisor, or subordinate partner for each language. The longlived Wynkyn de Worde, a native of Wörth in Alsace, was certainly in Caxton's employment at Westminster three or four years later, and printed English books for six decades, first as Caxton's foreman until his master's death in 1491, and thereafter independently until his own death in 1535, when he must have been in his eighties. It may well be that de Worde had already served in the Cologne press as a teenage apprentice, and was handed over to Caxton as a proficient youth by Veldener in 1473-4. Colard Mansion, as we shall see shortly, was certainly printing Caxton's French books, probably as a junior partner rather than an employee, in 1475, and hence cannot have learned his craft later than 1474; so it is likely that Mansion also, though perhaps a few months later than Wynkyn's arrival, was instructed on Caxton's behalf by Veldener.

On the old hypothesis it was necessary to suppose that Caxton's first type was made for him at Cologne in 1472, and that he brought it with him to Bruges towards the end of that year. It now seems likely that Veldener manufactured it at Louvain in 1473-4, as part of the process of installing and equipping Caxton's press. That the type was cut and cast by Veldener and no other can hardly be doubted. Veldener it was who produced Caxton's second and third types (for Veldener used these himself and supplied modified versions to other printers), and there is no other natural or possible candidate for the making of type 1. But the choice of style evidently came from the customer, for Caxton's type 1: 120B. is modelled on a Burgundian bookhand quite distinct from the Cologne modes which Veldener usually favoured.[1]

[1] Duff rather exaggerated when he claimed that Caxton type 1 significantly resembled Veldener's Cologne type, for the few letters which are at all similar are merely those which are common to both styles. Elsewhere Duff remarks (*Century* 24): 'A study of Caxton's first types clearly shows that they very strikingly resemble those of John of Westfalia who printed at Alost in 1473 and moved to Louvain in 1474'—an inexplicable and groundless error! De Westfalia's types are of Venice origin, and are entirely unrelated to any of Caxton's.

Early printers used for each class of text a type modelled on the manuscript hand in which scribes would normally write, and their readers would buy, the same kind of text. Latin books were usually written or printed in gothic, subdivided into the heavy, upright and angular style called *textura* for liturgical works, and the rounder, less formal *rotunda* or *fere-humanistica* for most other texts. Except in Italy, where roman script or type was favoured, vernacular literature was reproduced in national or regional varieties of the semi-cursive bookhand known as *bâtarde* or *bastarda*. This, as the name suggests, was a compromise hybrid between gothic script (in which each letter was formed separately) and the everyday cursive hand (in which each word was written continuously without lifting pen from paper, as most of us still write today). *Bâtarde* type can be distinguished at a glance from the prevalence of semi-cursive features, such as looped ascenders (in b, d, h, k, l, v, w, or some of these), groups of two or more letters tied in ligature, and sloping long s and f with tail below the line. So, as his programme at Bruges was to be devoted exclusively to courtly vernacular texts, Caxton (or Mansion advising him?) chose an ornate bâtarde used for such texts by highclass scribes everywhere in Burgundian Flanders, including Bruges. No doubt he showed Veldener for model a manuscript in this style, perhaps the very one from which he had translated the *Recuyell*, or even (a remote though attractive possibility) one by Mansion himself; but in any case the design of the type was dictated by Caxton, not by Veldener. The result is handsome, though not a complete aesthetic success. The lavish and non-functional height, breadth, and wide spacing of letters, the vast variety of alternative forms,[1] the rather untidy freedom of shape, line, thicks, and thins, all show Caxton's desire to reproduce the general appearance of a luxury manuscript, without regard for space-saving, reduction of paper costs, or economy in type.

So, towards the end of 1474 and the first months of 1475, Caxton produced the first book ever printed in the English language, the first by an English printer, and the first outside Germany in a bâtarde type.[2] As the distribution of paper and quiring shows, he had printed simultaneously on three presses, one for each of the three 'books' of the *Recuyell*. At a rate of about three leaves or six pages daily,[3] the 352 leaves of the *Recuyell* probably took (allowing for Sundays, holidays, short winter days, and mishaps) about five or six months to print. As the *Recuyell* survives to this day in upwards of

[1] The type comprises about 160 different sorts, including no fewer than 25 for long s in combination with other letters.

[2] The next bâtarde type was used a few months later, towards the end of 1475, by Pasquier Bonhomme in Paris.

[3] By the mid-1470s progressive printers were already printing at the rate which became standard throughout the centuries of the handpress, of one sheet of four folio pages per day, impressing two facing pages at each pull. But various lines of evidence suggest that Caxton, until 1480, set up and printed only one page at a time, in which case his daily rate per press would be only half this.

sixteen copies the edition was doubtless not a small one by the standards of the time. The provision of a second stock of paper suggests a decision soon before printing began to increase the number of copies, and the total may have been in the region of 400 or 500. Like many other early printers Caxton felt entitled to proclaim the astounding advantages of the new mechanical art over the scribe's penhand. 'I have practiced and learned at my great charge and dispense to ordain this book in print after the manner and form as ye may here see,' he wrote in his final epilogue, 'and is not written with pen and ink as other books be, to the end that every man may have them at once, for all the books of this story called the Recule [*sic*] of the histories of Troy thus emprinted as ye here see were begun in one day and also finished in one day'. This riddling expression is one of Caxton's little jokes. The production of the whole book must have taken several months; but it is true of course that every copy of each page would be printed within a single day. Similarly at Mainz Fust and Schoeffer in 1457 and Gutenberg himself in 1460 had boasted of working 'without help of pen', and Ulrich Han at Rome in 1469 had claimed to 'print more in a day than could be written in a year'.

Caxton's first customers ('diverse gentlemen and my friends') were no doubt found in Bruges and the Low Countries among the English retinue at Margaret's court, the English colony of merchants, and visiting English envoys. But it may be questioned whether these were enough to dispose of a whole edition, and it would be natural for him to export copies for sale by or through the London and provincial booksellers. If so, Caxton must have made his first contacts with the English home book trade in 1475, a year before his return to England. Some copies he doubtless presented to particular friends, or to royalty. A unique *Recuyell* in the Huntington Library at San Marino, California, contains a contemporary frontispiece engraving of Duchess Margaret in her throneroom—complete with wimpled ladies in waiting, deferential or gossiping courtiers, and a pet monkey—graciously accepting a book from a kneeling author. The same volume bears an ownership inscription: 'This book is mine Queen Elizabeth late wife unto the most noble King Edward the Fourth of whose both souls I beseech Almighty God take to his infinite mercy above'.[1] Perhaps this is the very copy Caxton gave the Queen; perhaps the engraving was based on a manuscript illumination in the special copy (now lost) which he surely presented to Duchess Margaret; perhaps the dapper little author (though the roundfaced features under his bobbed hair are indeterminate and unexpectedly youthful for Caxton in his middle fifties) is a rather flattering portrait of Caxton himself, and the only one now surviving; and no doubt the

---

[1] The inscription is not in the Queen's autograph, but written by 'Thomas Shukburghe the Younger', presumably a secretary who was putting her library in order after her widowhood in 1483; so it is quite possible that she possessed the book long before.

engraver was the same who produced nine engravings in a similar style for insertion in the Boccaccio, *De casibus virorum illustrium*, which Colard Mansion printed at Bruges a year later, in 1476.

Caxton produced his second book, the *Game of Chess*, immediately after the *Recuyell*, as is revealed by the overlapping of its paper supply with the later quires of the *Recuyell*. Presumably he began printing *Game of Chess* soon after finishing the translation on 31 March 1475, and took no more than a month or two, perhaps completing the printing in June, for this is a shorter work of only seventy-two printed leaves. Caxton's translation is from a French version made a little before 1350[1] by Jean de Vignay from the Latin *De ludo scaccorum* written about 1300 by the Genoese Dominican Jacobus de Cessolis. The names, shapes, positions and moves of the chess pieces are ingeniously allegorised as emblems of human society: the all-important king and queen are flanked by their trusty judges (here called 'alphins', Arabic for elephant, and later replaced by the modern chess bishop), knights, and commissioners (the rooks), symbolising law, nobility, and administration, while the humble pawns, of course, represent the middle and lower classes. The queen's rook's pawn does double duty as gambler and messenger, and the others from left to right act as policeman, taverner, physician, merchant, draper, notary, ending with the king's rook's pawn, who is a farm labourer. Caxton's text includes supplementary material from a rival French version made in 1347 by another Dominican, Jean Ferron; chess was first invented, says Ferron, by a Chaldean philosopher named Exerses, in the hope of reforming Nebuchadnezzar's son Evilmerodach, 'a jolly man without justice who did hew his father's body into three hundred pieces'.[2] The author abandons his chess allegory before the game begins, and diverges into medieval *exempla* (profane anecdotes with edifying morals) about the vices and virtues of the social callings; still, a tenuous thread of continuity of idea links his book with the chess ballet at the court of Queen Eleuterilida in the *Hypnerotomachia*, Rabelais's borrowing of this in his Fifth Book, and so with Pope's game of ombre in *Rape of the Lock* and Alice's journey in *Through the Looking-Glass*. The French text, like most of the books Caxton chose to translate, was immensely popular at the Burgundian court; Philip the Good had no fewer than seven manuscripts of it in his scattered ducal libraries, and Louis de Gruthuyse owned another at Bruges. Caxton's lifelong policy of enriching English literature with the hitherto untranslated best-sellers of Burgundy was now fully formed.

[1] Vignay dedicated his translation to Prince John of France, before his accession to the throne in 1350.

[2] Evilmerodach is in the Bible (II *Kings* 25.27), but the rest is apocryphal. Chess proved highly successful, and he became quite a reformed character. This composite Vignay-Ferron version was generally supposed to have been concocted by Caxton himself, but recent scholarship has revealed numerous manuscripts with a similarly conflated text, and there can be no doubt that Caxton used one of this class.

But the inner secret of Caxton's *Game of Chess* is only visible when we read between the lines of his dedication to George, Duke of Clarence, as no one seems to have done since Caxton's own day.

Ever since Barnet Clarence had replaced the slain Warwick as the chief focus of disruption in England. As Warwick's son-in-law, ally, and betrayer, false Clarence felt entitled to inherit all his vast estates; but cunning Gloucester arranged to marry Warwick's other daughter Anne Neville (widow of Edward the Lancastrian Prince of Wales) in order to claim half. In February 1472 Edward imposed an uneasy reconciliation on his brothers, giving Clarence the title of Earl of Warwick,[1] but awarding Warwick's northern domains to Gloucester, after keeping a few plums for himself. Clarence could never forget that Warwick and Louis had once promised to make him king. In 1473 he was implicated with Louis in the searaids on England and the occupation of St. Michael's Mount in Cornwall by the Earl of Oxford (a Warwick Lancastrian who had escaped from Barnet and lived to be Caxton's patron after Bosworth).[2] In November of that year, as the Pope's envoy reported, Edward was much exercised by the choice of governors to be left in charge of England during his coming invasion of France, 'so that he may not be overthrown by his brother Clarence'. In 1474–5 Clarence hired an Oxford don John Stacy to draw horoscopes predicting an early death for Edward and the little Prince of Wales.[3] If anything happened to them, Clarence would be King George of England! Meanwhile, like every other pretender or rebel of the century, from Bolingbroke to Perkin Warbeck (not forgetting Jack Cade, Edward's father York, Edward himself, Warwick, and Gloucester), Clarence let it be known that all England's woes came from the King's evil councillors, who ought to be replaced by his own followers and himself, as the King's loyallest and wisest subject. And this, just this, is the theme of Caxton's dedication!

Around the wording of Vignay's original address to Prince John of France[4] Caxton weaves a tissue, discreetly non-treasonable but transparent to any contemporary reader, of propaganda for Clarence and criticism of Edward's rule. 'I have understand and know,' he tells Clarence, 'that ye are inclined to the common weal of the king,

[1] Blades misidentifies the Earl of Warwick in Caxton's *Game of Chess* dedication of 1475 with the Kingmaker who died in 1471.

[2] Sir John Paston darkly reported that Clarence's and Rivers' men were buying arms in London, 'it is said for certain that the Duke of Clarence maketh himself big in that he can', and 'some think that under this there should be some other thing intended and some treason conspired'.

[3] Two of the alleged offences took place on 6 February and 20 May 1475, within the period when Caxton was writing and printing *Game of Chess*. Stacy and his accomplice Thomas Burdett, a landed gentleman in Clarence's household, were hanged at Tyburn on 14 May 1477 for these treasonable necromancies.

[4] Who became King of France in 1350, was captured by the Black Prince at the Battle of Poitiers in 1356, and died in captivity for want of ransom eight years later. Blades's statement that Vignay's and Caxton's words are the same '*nominis mutatis*' is as incorrect as his Latinity, but has deceived many.

his nobles, lords, and common people of his noble realm of England, and that ye saw gladly the inhabitants of the same informed in good, virtuous and honest manners, in which your noble person with guiding of your house aboundeth, giving light and example to all other . . . for God be thanked your excellent renown shineth as well in strange regions as within the realm of England.'[1] As for Edward, 'I pray Almighty God to save the King and to give him grace to issue as a king and to abound in all virtues and to be assisted with all his other lords in such wise that his noble realm of England may prosper and abound in virtues and that sin may be eschewed, justice kept, the realm defended, good men rewarded, malefactors punished, and the idle people to be put to labour, that he with the nobles of the realm may reign gloriously in conquering his rightful inheritance, that very peace and charity may endure in both his realms and that merchandise may have his course'. In short, Edward would 'abound in virtues' like Clarence himself if he made his brother a new Warwick, surrendered to his guidance, gave jobs to his noble friends ('good men rewarded'), and dismissed the lowborn Woodvilles ('malefactors punished').

Duchess Margaret's prompting is surely audible in this extraordinary manifesto. Caxton's allusion in *Recuyell* a few months before to the 'many good and great benefits' of which he hoped 'many more to receive of her Highness' suggests that she had promised to subsidise *Game of Chess* in return for his support of her favourite brother. But a note of personal grievance seems present in Caxton's complaint of Edward's failure to ensure 'that merchandise may have his course',[2] and in two remarkable insertions which he grafted into the *Game of Chess* text. One of these is a diatribe against English lawyers. 'Alas and in England what hurt do the advocates, men of law, and attorneys of court to the common people . . . how turn they the law and statutes at their pleasure, how eat they the people, how impoverish they the commonalty. I suppose that in all Christendom be not so many pleaders, attorneys and men of law as be in England only, for if they were numbered all that belong to the courts of the chancery, king's bench, common pleas, exchequer, receipt, and hell,[3] and the bagbearers of the same, it should amount to a great multitude. And how all these live, and of whom, if it should be uttered and told, it should not be believed.' As we shall see, Caxton himself was involved in a particularly vexatious Chancery suit at this time! The second insertion laments the old days when England 'knew

---

[1] A curious and perhaps not coincidental anticipation of an article in Edward's proceedings against his brother for treason in January 1478, alleging that Clarence had plotted 'to disinherit and destroy him by might to be gotten as well outward as inward'.

[2] An echo of the Warwickist grievance in 1470-1, that Edward had 'hurt merchandise'.

[3] This infernal nickname was given to the record office in the royal law courts at Westminster, because that was the place where dead documents were kept for ever.

justice and every man was in his office content' (like Caxton when Governor, and unlike his rivals?); 'how was renowned the noble realm of England, all the world dreaded it and spoke worship of it, how it standeth and in what abundance I report me to them that know it. If there be thieves within the realm or on the sea,[1] they know it that labour in the realm and sail on the sea . . . I pray God save that noble realm and send good, true and politic counsellors to the governors of the same.'[2]

In all this hankering for Clarence, this grudge against Edward and the established government, we seem to touch (though without much further enlightenment!) on the mysteries of Caxton's loss of his great office in 1470, and the further troubles which necessitated his application for pardon in March 1472, at the very moment when Edward imposed his settlement on the resentful Clarence, and sent his first postwar trade mission to Flanders. Even so, Caxton's rebellious sentiments did not go very deep. He was to understand soon enough that he had backed the wrong horses in Clarence and Margaret, and stake his future again on Edward, Rivers and Hastings.

In fact his loyal service to Edward had never ceased. At the time of the Hatcliff-Russell-Pickering negotiations at Bruges in March to July 1472, which failed to produce agreement either with Duke Charles or the Hanse, Caxton was in Cologne; but he may well have been consulted, or even made a brief appearance at Bruges in person, and perhaps needed his pardon to obtain clearance for this purpose. In the summer of 1473, when he had already returned to Bruges, Caxton is named among those commissioned to attend diets with Charles at Bruges and with the Hanse at Utrecht, at which the main articles of peace with the Hanse were drafted on 19 September, and agreed on 28 February 1474. On 10 August 1474 Caxton appeared in the refectory of the Carmelite monastery at Bruges (where the English merchants had their private chapel dedicated to St. Thomas à Becket, the mercers' patron saint) to witness the deposit of documents of ratification exchanged between Edward and the Hanse. Both sides had found fault with these documents, ostensibly because the seals were affixed with ribbons of parchment instead of silk!—but really as a pretext for another year of protracted haggling. Caxton acted again in the same affair in April

---

[1] Warwick, Fauconberg, the Hanse, Oxford, and the French had all been busy with piracy in the last five years. Had Caxton suffered personally? In any case, piracy was bad for trade.

[2] A third insertion strikes a happier note. 'I have myself been conversant,' says Caxton, 'in a religious house of White Friars [i.e. Carmelites] at Ghent, which have all thing in common among them and not one richer than another, in so much that if a man gave to one friar 3d or 4d to pray for him in his mass as soon as the mass is done he delivereth it to his superior or procurator, in which house be many virtuous and devout friars.' He may have lodged with the Carmelites when he visited Duchess Margaret at Ghent in March-June 1471 or later. Ghent was her usual residence in 1472-6. She also spent a few weeks at Bruges in January 1474—did Caxton take the opportunity to show her the early stages of the installation of his press?

1475, when a letter to the Bruges Hanse from the Hansards' secretary Herman Wanmate shows him specially concerned with the interests of the Hanse of Cologne. He was ordered by a royal warrant from Westminster dated 20 August 1475 to witness the final exchange of ratification documents (now duly rewritten and fitted with silk ribbons), and did so at Bruges in the Augustinian refectory on 4 September. Edward paid compensation of £10,000 to the Hanse (they had demanded £25,000), restored their privileges including the London Steelyard, and agreed to eject the rebellious Hanse of Cologne; but to his honour, and no doubt to Caxton's gratification, he delayed action in this last clause until the Hansards relented and readmitted Cologne in November 1476.

Meanwhile the Anglo-Burgundian trade negotiations still lingered on. Here, too, Edward commissioned Caxton on 1 December 1474 in a delegation which included the old familiar Yorkist faces, such as Hatcliff, Sir John Scott and Sir Thomas Montgomery, but was headed by a newcomer, the ex-Lancastrian churchman John Morton, the future inventor of Morton's Fork.[1] Charles seemed no more kindly disposed towards his victorious brother-in-law after Barnet and Tewkesbury; indeed, his inmost feelings were revealed in his top-secret declaration on 3 November 1471 that he, no other, was the rightful King of England! In fact, now Henry VI and the Prince were slain, Charles really was next in Lancastrian succession to the throne, being a great-grandson of John of Gaunt through his mother Isabella of Portugal, who died at Dijon on 17 December 1471. No doubt these congenital but phantasmagoric prospects were the true explanation of Charles's lifelong weakness for Lancastrians and lack of enthusiasm for Yorkists. His ambitions had now reached the point of megalomania. He planned to consolidate his scattered domains by further conquests from the North Sea to the Alps, to make himself King Charles of Burgundy instead of a mere duke and vassal of France, to induce the Emperor Frederick III to crown him, and in due course to become King of the Romans and Emperor himself. These eastward entanglements required that he should trick Edward into fighting Louis for him, while promising till the last moment to join in. So the real but hidden task of the 1474-5 mission to Flanders was to complete arrangements for Edward's expeditionary force against France.

This was Caxton's finest hour as a public servant. Charles gave his blessing and command for the provision of 500 ships from his dominions, and Caxton, as 'commissary and agent of the King of England', was charged with the raising and manning of the northern contingent of shipping from Holland and Zeeland. From 23 April to 26 May he visited the towns of Delft, Rotterdam, Gouda, Dordrecht, Middelburg and Flushing, escorted by a local dignitary Ghisbert van der Mye.

---

[1] Morton's Fork (the principle that if a man spends his money he can afford to pay heavy taxes, and if he saves it he can afford to pay even more) is still used to skewer the English taxpayer.

The rude sailors insisted on a month's pay in advance, but otherwise all went well. The town of Gouda stood Caxton and Ghisbert a dinner for 18s 8d. The invasion fleet began ferrying Edward's tremendous force of 13,000 archers, armoured cavalry and horrific artillery early in June, and Edward himself crossed to Calais on 4 July. Duchess Margaret called to see Edward on the 6th, and took Clarence and Gloucester back with her to Caxton's old haunt of Saint-Omer. Caxton had probably not set eyes on his new patron since May 1461, when Clarence was an eleven-year-old refugee at Bruges, and in his dedication he had modestly called himself 'your humble and unknown servant'. But it would not be surprising if Caxton seized this opportunity to ask Duchess Margaret for an introduction and to hand over the presentation copy of *Game of Chess* to Clarence in person. Perhaps Clarence even bought a few score to distribute among his lords and captains, with an offhand: 'I thought you might like this, he says some rather nice things about me.' The book was then at most a few weeks old, and in the epilogue Caxton had dropped heavy topical hints about the invasion already in progress, praying that Edward 'with the nobles of the realm' (which in the context meant Clarence and his gang) might 'reign gloriously in conquering his rightful inheritance', and wishing Clarence 'the accomplishment of your high, noble, joyous and virtuous desires'.

Ordinarily Clarence had few if any desires of that kind, but now Edward had let him out for once he was as good as gold. He brought a thousand archers (Gloucester had promised the same, but turned up with hundreds more to show off), and backed Edward when he decided to make terms with Louis (Gloucester opposed Edward for the first time, being all for war). Charles dropped in quite alone (half his armies were besieging Neuss in the territory of Cologne, the rest were looting in Lorraine), remarked annoyingly that with such a fine body of men Edward could march through to Rome if he pleased, and went away again. Edward, thus bilked by Charles, as he had expected and probably hoped, met Louis on 29 August at Picquigny (on a specially constructed bridge over the Somme, like Napoleon and Czar Alexander on the raft in the Niemen), and agreed to take his army home without fighting, in return for 75,000 crowns down and 50,000 annually as long as they both lived, with seven years' truce and free trade in France for English merchants. To the French, as sometimes happens, the mercenary conduct of the English seemed wellnigh un-English, but not to the English themselves. The break with Charles was total.[1] Everyone understood that the bold Duke was now friendless and doomed, and the star of

---

[1] According to that gossip-box the Milanese ambassador, Charles reacted by tearing his Garter to pieces with his teeth, and affirming that he, Charles, had a better claim to the throne of England than Edward, which was why Edward hated him so; he had a strong party in England, he said, and once he was King of England, 'he need only lift the other shoulder and forthwith he would be King of France'.

Burgundy was setting. Perhaps at this moment Caxton began to feel that Flanders was no longer a paradise for Englishmen, and turned his thoughts homeward.

So, in 1474–5, despite his semi-retirement, Caxton had the two busiest diplomatic years of his whole career, working simultaneously for trade agreement, Hanse treaty, and invasion of France. Yet this was the very period when he began his press, produced *Recuyell* and *Game of Chess*, and also translated the latter. How did he find time to print? The paradoxical but true answer is, that neither Caxton nor any other fifteenth-century publisher-printer would be at all likely to take any physical or manual part in the production of his books. Printing was the work of compositors and pressmen. If Caxton himself ever did a hand's turn at the press, or composed a few lines of type, it was doubtless in the style of the manager of the Grand Hotel at Balbec in Proust, on the day when 'with sacerdotal majesty', and with all his staff gathering round to admire, he 'carved the turkey poults himself'. *Game of Chess*, in particular, must have been already at press (if we recall the paper-evidence that it was printed without a break after *Recuyell*) at the very time in April-May 1475 when Caxton was absent in Holland and Zeeland collecting ships for Edward. The production of these two English books was no doubt superintended by Wynkyn de Worde, or by some unknown English-competent predecessor of Wynkyn. Caxton's own words, as the Hellingas point out, seem to disclaim direct collaboration in their printing, and perhaps even allude to his absence. Of *Recuyell* he says he has learned 'to ordain this said book in print', meaning apparently by 'ordain' that he made arrangements for its printing. As to *Game of Chess*, he notifies still more explicitly, in the new prologue to the second edition of about 1482, that the French original came into his hands 'at such time as I was resident in Bruges in the County of Flanders', and that after translating it he 'did do [i.e. caused or ordered to be] set in imprint a certain number of them, which anon were depeshed[1] and sold'. The same significant distinction was preserved at his Westminster press. In books produced during the first ten years he says outright 'emprinted by me William Caxton', no doubt because he was then actively supervising his press on the spot.[2] But in the last six years up to his death in 1491, beginning with his edition of Malory's *Morte d'Arthur*, 31 July 1485, he usually preferred the Latin phrase '*Caxton me fieri fecit*' (Caxton caused me to be made, the exact equivalent of 'did do'), presumably because in advancing old age he had again handed over the daily management of production to Wynkyn de Worde.

[1] A nonce word from *dépêcher*, like Recuyell from *Recueil*. Caxton followed the Red Queen's advice to Alice, 'to speak in French when you can't think of the English for a thing'.

[2] However, in the prologue to his second edition of *Canterbury Tales*, of 1483, when describing the printing of his first edition, in 1478, Caxton uses almost the same words as in *Game of Chess II*: 'I did do emprint a certain number of them, which anon were sold.'

Another of Caxton's preoccupations at this time was the singularly vexatious and undeserved lawsuit in Chancery which goaded him a few months later to his diatribe against English lawyers, with special mention of the Court of Chancery, in *Game of Chess*. In November 1474 he had written to London from Flanders to authorise repayment of a 'debt'[1] of £190, which he owed to his old acquaintance John Neve, mercer, from goods and money held on his behalf by another mercer, John Salford. This is the same Neve who had partnered Caxton in a £200 deal in linen at Ghent twenty years before, in 1455. Salford paid only £28 on account, and persuaded Neve to wait till Christmas for the rest, so that he might sell Caxton's wares to better advantage. But alas, instead of keeping his word Salford next sued Caxton before the Lord Mayor's court for £200 ('where in deed the said William Caxton owed him then no penny, nor yet doth,' as Neve affirmed), attached his goods, and 'hung still as a dormant to the intent to defraud your said suppliant'. Neve had no recourse but to counter-attach the goods held by Salford and sue the innocent Caxton for the outstanding £162. Finally Neve filed a plea on 28 January 1475 for the dispute to be settled 'according to faith and good conscience' in the King's Court of Chancery.[2] The outcome is unknown; let us hope all parties recovered their dues. But the case is interesting, both as accounting for the hitherto unexplained allusion in *Game of Chess*, and as showing that Caxton, even when his press was in full swing, was still in business as a merchant for a large sum. A last insertion in *Game of Chess* is more ambiguous. 'Alas,' says Caxton, 'how keep the princes their promises in these days, not only their promises but their oaths, their seals and writings and signs of their proper hands all faileth, God amend it!' Some might read it as a rebuke for the perfidy of Duke Charles, who had left Edward to fight Louis alone; but to Clarence (himself second to none for treachery) the cap might seem to fit Edward, for awarding Warwick's northern lands to his brother Gloucester.

[1] Presumably, as usual, not a debt in the modern sense, but a sum due under a credit transaction.

[2] The year is not specified in Neve's plea. However, he refers to the Chancellor as 'Bishop of Lincoln', which can only mean Thomas Rotheram, who was Bishop of Lincoln in 1472-80 and became Chancellor in May 1474. So the January in question must be either 1475 or 1476, because from September 1476 onwards Caxton was resident in Westminster; and the day of the week is given as Saturday, which fits only 28 January 1475.

9

PRINTING AT BRUGES II:
COLARD MANSION

IMMEDIATELY after completing *Game of Chess* towards midsummer 1475 Caxton produced a series of four books in French, which must have fully occupied the following fourteen months until his move to England in September 1476, for they totalled 520 folio leaves. Why in French? Perhaps, as Blake suggests, Caxton found the marketing of English books at long distance was impracticable. Then again, he had no third English translation ready, and no prospect of leisure to write one. But probably the chief cause both of his recourse to French texts, and of his ability to continue printing without interruption at the height of his diplomatic duties and absences in the summer of 1475, was the planned presence in his firm of the Bruges calligrapher Colard Mansion.

Colard Mansion makes his first appearance in 1450, when Duke Philip the Good's keeper of jewels Guillaume de Poupet paid him the large sum of 54 livres for a manuscript on vellum bound in blue velvet of the *Romuleon* (a history of Rome from Aeneas to Constantine),[1] which was shelved in the palace library at Bruges. He became a founder member of the Guild of St. John the Evangelist, a confraternity of the Bruges booksellers, scribes and allied trades, when Duke Philip granted it a charter in 1454, and rose to be Dean of the Guild from 1 January 1472 to 31 December 1473. Mansion paid his annual dues regularly from 1454 to 1473, missed payment in 1474 (perhaps implying absence?), paid again in 1475, and then ceased (perhaps implying change of occupation?) until a single and final payment in 1483.[2] Mansion began activity as an independent printer in 1476, when he produced a Boccaccio, *De casibus virorum illustrium* in French, some copies of which are illustrated with a series of nine masterly copperplate engravings, the earliest in any printed book, and evidently by the same artist (perhaps the Master of Mary of Burgundy) as the engraving of Caxton in Queen Elizabeth Woodville's copy of *Recuyell*. This Boccaccio was preceded by a Pierre d'Ailly, *Jardin de dévotion*, with a colophon reading: *Primum opus impressum per Colardum Mansion. Brugis laudetur omnipotens.* (The

---

[1] In a French translation made for Philip the Good by Jean Miélot from the Latin original written in 1361-4 by Roberto della Porta of Bologna.

[2] As Sheppard points out, Van Praet and Crotch misinterpret the Guild records in assigning Mansion's decanate to 1471-2, and in supposing that his dues for 1473 were paid on his behalf by the Guild clerk Joris Caelwaert. The latter entry in fact records Caelwaert's payment of his own dues 'for Colard Mansion's year [i.e. as dean] which fell in 1473'.

first work printed by Colard Mansion. At Bruges the Almighty be praised). During the next eight years he produced a total of twenty-four editions, mostly in French,[1] ending in May 1484 with a huge and magnificent French Ovid, *Métamorphoses moralisées*, which evidently overstrained his finances, for he fled in debt, leaving the rent of his shop in the cloisters of St. Donatian unpaid,[2] and is heard of no more. Mansion's first type is a lavishly enlarged version of Caxton's second, and was certainly made by Veldener.

It is tantalising that Caxton and Mansion are never mentioned together in any document, but nonetheless obvious that they must have been linked in some way. The precise nature of their business connection has exercised scholars for more than a century, with dramatic reversals of opinion when new evidence or arguments appear. Blades in 1861 maintained that Mansion taught Caxton to print, and himself printed the two English books with Caxton as pupil, and the four French books alone. L. A. Sheppard in 1952 showed that Mansion's own press did not commence until several years after Caxton's, and hence concluded that it was Caxton who taught Mansion, and that Caxton alone was responsible for the printing of all these books. Wytze and Lotte Hellinga turned the tables yet again with a new and convincing interpretation of Mansion's colophon to *Jardin de dévotion*. In this highly unusual and significantly emphatic tailpiece Mansion seems to declare, covertly but unmistakably, that *Jardin* is the first book which he has printed in his own name and on his own account as an independent printer-publisher, in contrast to other books which he has printed anonymously and on behalf of another person. If this is so, these books were surely the four French editions, and the unnamed associate must have been Caxton himself.

But the strongest and most direct evidence of Caxton's and Mansion's association is to be found in a comparison of the characteristics of each as producer of texts and printer or scribe, and in the significant similarities which this reveals. Caxton's lifelong practice of creating hitherto unpublished material for his press by writing new translations in his own language is unique among fifteenth-century printers, with the exception of Mansion himself. Mansion's way was the same, both as scribe and as printer. At least four of the French texts which he and the employees in his Bruges scriptorium multiplied in manuscript were Mansion's own translations from Latin originals, and two of these he later printed. Equally unusual in a printer is Caxton's fancy for supplying prologues and epilogues to his publications, or adapting those of the original author, with the addition of personal anecdotes, fragments of autobiography, and comments on the times. Mansion did likewise in his four

---

[1] Twenty-one in French, one in French and Latin, and two in Latin.

[2] Mansion took his shop in the cloisters under the dormitory in 1478 for six years at 32 shillings a year. Blades misrenders *ambitus* (cloisters) as 'church porch'.

translations,[1] and inserted vivid lamentations on the calamities of the year 1477 in his edition of Boethius, *De consolatione philosophiae* in French and Alain Chartier's *Quadrilogue*. Caxton's editorial activities are no less unwonted: he takes pains, pride, and pleasure in collating, adapting, improving, or supplementing his text (as in *Dicts*, *Canterbury Tales II*), discussing its differences from other authorities (as in *History of Jason*), compiling new material from other sources (as in *Polycronicon* and *Golden Legend*), inserting everywhere his own remarks or reminiscences. Such too was Mansion's habit, notably in *Donat spirituel*, where he rewrites Gerson's original in the form of a dialogue, and in the *Ovid*, where he creates a complexly unique text full of additions both from other sources and of his own invention.[2] Caxton's recourse to royal or noble patronage, which he began at Bruges with Duchess Margaret and Clarence and continued still more extensively at Westminster, was also characteristic of Mansion, who worked under the patronage of Duke Philip and of Louis de Gruthuyse.[3]

Such similarities of practice, so close, so detailed, and so exceptional, can hardly be due to coincidence. They demonstrate not merely a personal connection between Caxton and Mansion, but a profound and eager exchange of ideas and ambitions. Caxton must surely have modelled himself on Mansion as early as 1469-71, when he set himself to translate a Burgundian bestseller, to embellish it with prologues, and to obtain the patronage of Duchess Margaret. Mansion had then been a prominent scribe at Bruges, Caxton a wellknown resident foreigner, ever since the late 1440s, and they may well have been already acquainted for many years. When Caxton went to Cologne in 1471 the thought of printing was no doubt far from their minds; Mansion can have had no part in the decision to learn the new alien art of printing, which came from Caxton's sight of the Cologne presses and his meeting with Veldener. But when Caxton returned to Bruges as a trained printer in 1473, and proceeded with Veldener's aid to set up his own press, what was his further contact with Mansion?

This problem, once again, can be elucidated by a comparison of the methods of the persons concerned. Strangely enough, the contrast in style and technique between Caxton's work and his master Veldener's, particularly as exemplified in the great *Bartholomaeus* on

[1] These included the popular book of animal fables *Dialogus creaturarum*, a *Pénitence d'Adam*, Gerson's *Donatus moralisatus*, and Ovid, *Metamorphoses moralisatae*. Mansion printed the two last.

[2] In the second of his *Ovid* prologues Mansion tells the story, very much in Caxton's manner, of how he witnessed the felling of an oak at the monastery of St. Michael in Antwerp.

[3] Mansion translated *Pénitence d'Adam*, an apocryphal narrative of the Fall, at Gruthuyse's command, and supplied him with a copy of his *Dialogue des créatures*. He calls Gruthuyse his *compère*, and himself Gruthuyse's, perhaps meaning that they had stood as godfather to one another's children, but more probably using the word to mean 'friend'.

which Caxton learned his craft, is as striking as the resemblance between Caxton's and Mansion's. This contrast is not a matter of proficiency, as between an accomplished master-printer and an imperfect pupil (for Caxton and Mansion show little inferiority in quality to Veldener), but of the fundamental antithesis between two cultural ideals. Veldener's type and lay-out display the severity, formality, and economy of scholastic Cologne, Caxton's and Mansion's the luxury, fantasy and lavishness of the Burgundian manuscript. The style of the six books from Caxton's Bruges press was continued throughout the duration of his Westminster press, for the first six years without noticeable change, thereafter with some evolution towards economy and modernisation, but always retaining its Burgundian inspiration. Mansion, when he began his own press in 1476, printed recognisably in the same style as Caxton's,[1] but evolved in the other direction, with increased lavishness and magnificence.

The technical connection between Caxton and Mansion is most signally shown by two features, the dogmatic irregularity of line-ends and the unique method of redprinting. The older practice of allowing uneven line-endings, which many printers had adopted from scribal tradition, was everywhere being replaced in the early 1470s by the newfangled way of spacing out each line so as to present a straight right-hand edge to the type-page. The anonymous Cologne press did so early in 1472, and the new method is already found in the *Bartholomaeus*; so Caxton must have learned it from Veldener in 1472. Yet all six books from his Bruges press have uneven line-endings.[2] Mansion at his own press did not change to even endings until 1479, after printing six more books in the old manner. Caxton at Westminster went modern later still, in 1480. Uneven endings are usually considered a mere symptom of lack of skill or care, and it is true that they occasionally occur in a good printer's earliest works,[3] or in the whole output of a minor printer,[4] even in the late 1470s. But they are primarily an intentional assimilation of normal scribal practice, and tend to be retained longer than usual by printers who began as scribes, or (as in the case of Gutenberg and Schoeffer) employed an ex-scribe for matters of layout.[5] Their appearance in the first six books at Bruges, and their late persistence in Caxton's and Mansion's

[1] As the Hellingas remark (HPT 23): 'There are marked similarities of technique and layout in the work of the two firms.'

[2] Not consistently. Exceptions (including a few pages in *Recuyell*, many in *Recueil*, *Méditations*, and most in *Jason*) suggest that at least one compositor was unable to unlearn previous training in even endings. *Game of Chess* and *Cordiale* are uneven throughout.

[3] For example, Gerard Leeu's first book at Gouda, an *Epistles and Gospels* in Dutch, 1477.

[4] For example, in all the books of the Printer of *Haneron*, an anonymous unlocated press in Holland, active in 1476-7.

[5] Schoeffer used uneven endings in many of his most magnificent books up to 1473. Gutenberg himself (if I am right in identifying him with the printer in the 36-line Bible type) adopted even endings as early as 1457.

subsequent work, belongs to the tradition of the Burgundian calligraphic manuscript, and must surely be attributed to Mansion's guidance.

A unique feature of Mansion's work is the extraordinary method of printing in red. The normal practice from earliest times was to print red and black from separate inkings and pulls of the press.[1] Mansion's procedure throughout his separate career[2] was to ink the whole page in black, to wipe off the ink from the appropriate portion, re-ink this in red, and then print off at a single pull. This process is unsatisfactory, as the red ink is visibly soiled by the underlying remains of black, and adjacent areas of black print are touched with red.[3] Here again the closest parallel is found in the ex-scribe Peter Schoeffer, who produced the splendid two-colour printing in his 1457 and 1459 Mainz Psalters by a single-pull method, but avoided Mansion's ugly effect by inserting the red type ready inked. No direct contact between Mansion or Schoeffer need be envisaged; but the process, especially in the detrimental form used by Mansion, is one which could only be devised by a scribe, and is inappropriate to a printer with an exclusively typographical background. Veldener himself was familiar with the normal two-pull method, which he used at Cologne and Louvain in 1473.[4] Mansion's faulty method is already found at Caxton's Bruges press in the French *Cordiale*.[5]

Caxton's first type, as we have seen, was no doubt cut and cast by Veldener, but is of Burgundian style. It bears a distinct resemblance to the hand or hands in the manuscripts of Mansion's *Pénitence d'Adam*[6] and *Dialogue des créatures*. Whether these are in Mansion's autograph is uncertain; if not, they may well be by employees in his scriptorium trained in his style.[7] It seems likely that Mansion him-

[1] Gutenberg himself did so, in the few lines of red printing which occur in the 42-line Bible (c. 1453-4).

[2] As in *Boccaccio*, 1476, *Boethius*, 1477, *Somme rurale*, 1479, and *Ovid*, 1484.

[3] Blades wrongly suggests that the red type was re-inked with the finger; but the reddening of nearby black type, despite the fact that the red type are intentionally spaced as far as possible from these, shows that the inkball was used.

[4] At Cologne in *Flores sancti Augustini*, at Louvain in Boccaccio, *Genealogia deorum*. Rather significantly, the only other printer to use red in Cologne at this time was ther Hoernen, who may have been Veldener's pupil. Ther Hoernen, on the other hand, persisted in uneven line-endings rather late, until 1475.

[5] The *Recuyell* has a single page (the first) printed in red; but as this page contains no black, the question of method does not arise.

[6] Blades pointed this out in the *Pénitence*, being unaware of the *Dialogue*. His remark has been unduly neglected, because he used it to support his mistaken view that Mansion made Caxton's types and taught him to print; but it seems none the less valid as to fact.

[7] Mansion's apprentice is mentioned in St. John's Guild records in 1458, and his assistant Hannekin in 1462 and (as dead) in 1483. The gender shows he was male, and not as Crotch supposes a woman! On 1 December 1480 Mansion contracted to produce a manuscript of Valerius Maximus in French, 'in his own hand or as good', for the former chamberlain and army general of Charles the Bold, Philippe de Hornes. (His requests for payment by advance instalments do not indicate poverty, as Blades supposes, as this was a normal

self designed Caxton's first type for Veldener to manufacture. The same is true of Caxton's second type, used at Bruges in the *Cordiale*, of which Mansion used an enlarged version at his own press.

This strong and abundant circumstantial evidence seems to show that Mansion played a major part in Caxton's Bruges press from the very beginning. As early as 1469-71, when they were probably already old acquaintances, Caxton must have discussed with him the choice of text, patronage, and editorial presentation of his *Recuyell* translation, which are founded on Mansion's methods. Caxton reached Cologne from Ghent in June 1471, still as it appears with no thought of printing, and made his great resolve towards the end of the year; he returned to Bruges early in 1473 a trained printer, began with Veldener's help to set up his press, and no doubt revealed his plans to Mansion. Mansion, as a thriving master-scribe in the Bruges manuscript trade, would be an ideal choice for the design of his type and layout and for the marketing of his products. Their partnership probably began early in 1474, immediately after the close of Mansion's 1472-3 decanate in the Guild of St. John. Mansion's disappearance from the records of the Guild throughout the period 1474-82 (except for a last isolated payment of dues in 1475) no doubt resulted from his change of occupation. During the first half of 1474 he will have designed Caxton's type 1 for Veldener to cut and cast, and have learned printing himself, presumably from Veldener at Louvain or during visits made by Veldener to Bruges. During Caxton's absences in 1474-5 Mansion may well have lent a hand with the production of *Recuyell* and *Game of Chess* (though not with the composition or proofreading, which would require an Englishman or English-speaker). He was no doubt chiefly responsible in 1475-6 for the four French books. Possibly his contract with Caxton provided that Mansion should undertake these books, with a share in their marketing and profits, and thereafter be free to commence his own press. In any case, Mansion cannot have been brought in as a mere stopgap after *Game of Chess*, in response to an unforeseen changeover from English production to French, as this would not leave him sufficient time to learn printing. Mansion's entry into the firm, followed by his training, must have occurred well before the *Recuyell* went to press in the autumn of 1474. However, the four French books no doubt mark the period, from summer 1475 to summer 1476, when he became the working executive partner in the firm of Caxton and Mansion.

As we have seen, the French books must have followed without intermission after *Game of Chess*, since otherwise there would be too little time for this bulky new output of 520 leaves within the fourteen months before Caxton's departure to England. None contains a date.

---

procedure with scribes, and even with printers, when working on a lengthy commission.) In 1482 he completed a *Dialogue des créatures* manuscript in his own translation for Louis de Gruthuyse.

The first was probably Raoul Lefèvre's *Recueil des histoires de Troie*.[1] Apparently three presses were worked simultaneously as with *Recuyell*. Five quires from the end of *Recueil* the third press ran out of the main stock of paper (watermarked with the fleur-de-lis shield of France) and brought in a mixed batch (partly with a unicorn watermark from Troyes in Champagne, partly with an ox). The same mixed paper was also used in Pierre d'Ailly, *Méditations sur les Sept Psaumes pénitentiaux*, which must therefore have been the next to be printed. Irregular line-endings tend to occur on the versos in the last section of *Recueil* and in *Méditations*, which shows that two compositors of different habits were serving the same press, and that only one page was being printed at a time. These were the first books ever printed in French, just as *Recuyell* was the first in English.[2]

Cardinal Pierre d'Ailly (1350-1420), the voluminous author of the *Méditations*, is still remembered today for a book which helped to convince Columbus that the world was round,[3] and also, less agreeably, for having condemned the Czech martyr John Huss to the stake in 1415. Some of his 174 known works circulated particularly in Burgundian domains, perhaps because he was bishop of Cambrai, a diocese which then (much to the indignation of the Duke at that time, Philip the Hardy) encroached as far as Brussels and Malines. Mansion was soon to print his *Jardin de dévotion*, possibly without knowing his authorship, for Mansion's edition does not contain the Cardinal's name. The Seven Penitential Psalms[4] were standard devotional reading for lay folk, both separately and as part of the *Book of Hours*. D'Ailly's meditations on them had already been printed in the Latin original by ther Hoernen at Cologne about 1472 (as Caxton must have been well aware), and the anonymous French translation belonged, like the other books from the Bruges press, to the favourite reading of the Burgundian court.

Raoul Lefèvre's Golden Fleece romance *Histoire de Jason* was last of the five books in type 1. This French edition at Bruges must be connected in some way with Caxton's English translation, which became one of the first books from his Westminster press, early in 1477. Perhaps he had planned to follow up *Game of Chess* with an English *Jason*, but was prevented by his rush of diplomatic business in the summer of 1475; or perhaps it was his French edition that gave him the idea of producing a translation. In any case, he probably

[1] Manuscripts of this work were already numerous, and it cannot be assumed that the same one which Caxton had used for his translation six years before served also as copytext for the printed edition.

[2] The first French texts printed in France were a *Legenda aurea* from Guillaume Le Roy at Lyons, 18 April 1476, and a *Chroniques de France* from Pasquier Bonhomme in Paris, 16 January 1477.

[3] A copy in the Biblioteca Columbina at Seville of the first edition of d'Ailly's *Imago mundi*, printed about 1483 by Johannes de Westfalia at Louvain, contains woodcut diagrams of the spherical earth, and is profusely annotated by Columbus himself.

[4] *Psalms* 6 ('*O Lord, rebuke me not in thine anger...*'), 32, 38, 51, 101, 130, 143.

worked on this English translation before leaving Bruges, as he would hardly have had time for it later.

Last came another Burgundian favourite, the *Cordiale*, a horrific treatise on the Four Last Things (death, judgement, hell, and heaven), written towards 1380-96 by Gerard van Vliederhoven, an official in the monastery of the Teutonic Order at Utrecht, and translated from Latin into French in 1455 by Philip the Good's secretary Jean Miélot. This book is a star witness to the collaboration of Caxton and Mansion, for it shows Mansion's characteristic redprinting, and is printed, evidently as a trial piece, in the new type 2: 135B. which Caxton took to Westminster soon after, and which none but Mansion can have designed for him. Comparison of individual letters and of massed effect on the printed page shows the reasons for abandoning type 1 and the improvements expected from the new fount. Both are based on typical and related Burgundian manuscript hands (possibly Mansion himself used each in his repertoire as a scribe), and many similar letter-shapes show their common origin. But type 1, with its broad, rounded, widely spaced letters, is sprawling and lacks unity of design. Type 2 illustrates a stage in the evolution from calligraphic imitation to typographical autonomy which in the 1470s was taking place everywhere in European printing. Formal splendour has replaced the caprice and liberty of handwriting; the letters are narrower and taller, angular and spiky; the curvilinear element is retained and exaggerated by means of head-to-tail leftward flourishes on many capitals (A, B, M, N, P, R, V). The impression of lavishness is produced vertically instead of horizontally; the height of 20 lines is increased from 120 to 135 mm.[1] The total effect of selfconsciously frivolous solemnity belongs like tall felt hats and winklepicker shoes to the brief high summer of Burgundian culture; ironically, for the type lasted longer than Charles's Burgundy, and Burgundy style lived on in England as Caxton style.

Type 2 was certainly cut and cast by Veldener, who used it himself at Louvain in 1475 in an early, experimental version as a heading type (that is, a display type for page and chapter headings) for a law text, Angelus de Gambilionibus, *Lectura super titulo de actionibus*. So, if we are right in thinking that Caxton ordered type 2 for use in England, then he must have planned his departure as early as the summer of 1475, a year ahead. Colard Mansion must have designed his own first type not long after and ordered it from Veldener, as it was ready for use towards the summer of 1476. Mansion 162B. is closely modelled on Caxton's type 2, but goes one better still in luxury and magnificence, for it is much enlarged both in body height and in size of face. The brilliant career of type 2 continued for many

---

[1] Type 2, being taller, gives only 28 lines to the page (though Caxton increased this to 29 lines at Westminster) instead of type 1's 31 lines, but being also narrower can accommodate one more word in each line, with the no doubt intentional result that the two types produce practically the same page contents.

years. Veldener made a reduced and simplified version for his own use in Dutch texts when he moved to Utrecht in 1478, and supplied other variants to Jean Brito, Mansion's rival at Bruges, in 1477, and to William de Machlinia, Caxton's rival in London, in 1484. Caxton's own types 4 (1480) and 6 (1488)[1] were designed on the same model, and must surely have been made to his order by Veldener.

Caxton intended his type 2 as a bâtarde text type for use with the vernacular, and still required another type, a gothic, for headings and Latin. His type 3: 135G. was in use in his 1476 *Indulgence* immediately after his arrival at Westminster, and must therefore have been ready before he left Bruges. This is a vertical, angular gothic in the international style called textura (from the interwoven effect caused by its equidistant uprights) with face larger than type 2, but with body of the same height, for convenience in conjunct setting. Veldener, yet again, was the maker, for he used it eight years later, after his return to Louvain in 1484, and supplied it to other printers. No one has yet assigned the design of Caxton's type 3 to Colard Mansion; but when we notice that many of the capitals (B, E, F, H, O, Q, T) are the same as in type 2, and that most of those which are different (C, D, G, M) are as in Mansion's 162B., we can only ask: who else?

When Caxton and Mansion liquidated their partnership in 1476 they apparently divided unsold stock between them. A French *Jason* in the Bibliothèque Nationale, Paris, is bound with Mansion's *Quadrilogue* of about 1478, suggesting that the two were sold together. A *Méditations* and French *Cordiale* in the British Library are bound together and contain notes by the same contemporary English owner; perhaps Caxton took these to England and sold them there, though they might of course have been bought at Bruges by an English visitor. A manuscript in Westminster Abbey Library[2] contains, as a relic from a pastedown of the original binding, a scrap of *Recueil*, printed on one side only, either by mistake or as a trial pull; so Caxton must have been thrifty enough to take press waste from Bruges to use at Westminster. A rather surprising number of books, surely too many for coincidence, were printed both by Mansion at Bruges in French and by Caxton at Westminster in English;[3] so perhaps they continued to exchange ideas after parting.

The first dated book in Caxton's type 6 is of 1489; but I think it likely on various grounds that he acquired it in 1486, the year of Veldener's retirement from business, and used it in 1488 in undated books.

[2] Abbot John Flete's history of the Abbey, copied by the scribe William Ebesham, a resident of the Abbey throughout Caxton's period and a servant of the Paston family, of whom more later, and bound by the Caxton binder.

[3] *Boethius, Cato, Dicts, Art of Dying, Controversie de noblesse, Ovid.* Caxton was first with *Dicts* and *Ovid*, Mansion with the rest. Neither used the actual edition printed by the other, so the contact, if any, was of ideas only. Was Mansion also linked, perhaps through Veldener, with Veldener's former partner Johann Schilling? Schilling at Vienne in Dauphiny and Mansion at Bruges each printed Alain Chartier's *Quadrilogue* towards 1478, and Peter Schenck, Schilling's successor at Vienne, produced in 1484 an *Abusé en cour*, a work previously printed by Mansion.

Caxton's Bruges press must have ended towards midsummer 1476, in order to leave time for his installation at Westminster by Michaelmas, and for Mansion to produce *Jardin de dévotion* (30 leaves) and Boccaccio, *De casibus* (292 leaves) before the end of the year.[1] Perhaps he had envisaged his move to England all along; he could not wish to print English or French books at Bruges for the rest of his life, and his Bruges press was no more than an experimental preliminary for his return home. But the ominous isolation of Charles the Bold in summer 1475, which Caxton witnessed at such close quarters, may well have warned him to leave while the going was good. If so, the events of 1476 confirmed his foresight. Charles was madly fighting the Swiss, who broke the Burgundian army at Grandson on 2 March and Morat on 22 June, and was obviously doomed. The Burgundian Netherlands were already faced with the menace of French invasion and internal disruption, which came true in 1477, and (as a minor calamity among so many) halted printing at Bruges, Louvain, and Brussels. Meanwhile Charles was still in touch with Clarence, and announcing his intention to become King of England. Edward did not know whether to be glad or sorry for his brother-in-law's difficulties, but thought the moment ripe to try again for the removal of the cloth embargo. On 12 September 1476 he commissioned a trade delegation (Hatcliff, Scott, and other old hands) to go to Bruges, at the very time when Caxton was leaving for England to keep his rendezvous with our quincentenary.

[1] It is possible that *Boccaccio* overlapped into 1477 by Flanders dating, but not likely, for Mansion had completed his equally long and splendid *Boethius* by 28 June 1477, besides at least one other work, *Controversie de noblesse* (40 leaves), which belongs before the *Boethius* on typographical grounds.

10

A SHOP IN THE ABBEY

O N 30 September 1476 Caxton paid ten shillings to the sacrist and abbot John Eastney[1] as a year's rent for a shop (*'una shopa'*) in the precincts of Westminster Abbey. When a memorial plaque to Caxton was erected in 1954 the Abbey Librarian L. E. Tanner identified the exact site by following up later entries for the same shop in the sacrist's rent rolls. Caxton rented these premises every year until his death in 1491, and was succeeded by his assistant Wynkyn de Worde until 1500, when Wynkyn moved to Fleet Street in the City. In later years (when it was occupied by other members of the book trade, beginning with a certain James Bookbinder) the shop was described variously as 'adjoining the Chapter House', and 'near the south door of the Church'. This places it between the two northernmost of the Chapter House's flying buttresses, a few paces to the right as one leaves the south transept by the Poets' Corner door. The place was singularly well chosen, on the path leading from the King's Palace of Westminster (where the Houses of Parliament now stand), which housed the royal residence, the court, the seat of government, the Lord Chancellor and Council, the chief lawcourts of the realm in Westminster Hall, and the House of Lords, to the Abbey Church where the dignitaries from these thronged to mass (as an alternative to their own chapel of St. Stephen's) or to take the shortest cut through the precincts. The Chapter House itself in Caxton's time was the meeting place of the House of Commons,[2] and the sacrist's accounts show that the shops next door to Caxton's, though usually empty, were sometimes let specially 'in the time of Parliament'.[3] Many merchants found it necessary to visit Westminster or convenient to reside there, and the City itself was only two miles away by Strand or river. So this prudent printer for royalty, nobility, law, church and middle classes had picked the supreme centre and thoroughfare of all these.

[1] Eastney was elected Abbot in August 1474, when his predecessor Thomas Milling was made Bishop of Hereford, but chose to keep his key posts as sacrist and warden of the fabric works.

[2] As it happened Parliament had recently ended a long session from 1472 to 1475, and was to sit only once (January-February 1478, to liquidate Clarence) until just before Edward's death in 1483. It met in 1484 under Richard III, and three times under Henry VII, in 1485, 1487, and 1489-90, before Caxton's death in 1491.

[3] Notably during Henry VII's third Parliament, at 4d for a week and 2s for the whole session. The assumption of Crotch, Tanner and others that these lettings were to Caxton himself is not justified. I am indebted to Mr. Howard M. Nixon, Librarian of Westminster Abbey, for confirming this view, which he had already formed independently from inspection of the original document.

But several other reasons besides this unique concourse of customers must have influenced Caxton's choice of Westminster. One of these was the presence, as we have already noticed, of other Caxtons, including William Caxton who was buried at St. Margaret's in 1478 (he may have been the Printer's father, if Thomas Caxton I of Tenterden wasn't),[1] John Caxton who in 1474-7 joined the same church's lay guild of Our Lady of the Assumption, and above all Richard Caxton, a monk of the Abbey from 1473 to 1504. Perhaps Richard bespoke the Chapter House shop for the Printer before he left Bruges, and recommended him to the Abbot-sacrist Eastney. Then again, although Caxton still kept lifelong friends among his brother mercers in the City, he had good reason to prefer the protection of Court, Law and Church at Westminster to the chauvinism of unruly London, where the guild of Stationers was as yet restricted to the manuscript trade, and the foreign craftsmen of the printer Richard Pynson, a Frenchman from Normandy, were assaulted as late as 1500. Caxton's first workmen on arrival from Bruges must have included other foreigners besides the Alsatian Wynkyn de Worde, and it was not till January 1484, in the reign of Richard III, that Parliament provided expressly for the protection and residence of alien printers and booksellers in England.

Perhaps it was to insure himself against such risks in the first months of his return that Caxton ('late of London, mercer') took out letters of protection under the King's privy seal on 2 December 1476, valid for one year, as a member of the retinue of his former colleague Lord Hastings, lieutenant of Calais, ostensibly 'for the safe custody, victualling and defence of the said town and marches'. Perhaps Caxton really did help Hastings in some matter of shipping,[2] as he had Edward himself in 1475; or perhaps he needed a cover for importing his press equipment and foreign pressmen by sea from Bruges. But probably his chief motive was to get immunity during the year of his homecoming from the chicanery of business rivalries, or from political difficulties of the kind that had necessitated his 1472 pardon.

Caxton's first datable piece of printing in England is an *Indulgence* issued by Pope Sixtus IV in aid of the war against the Turks, of which the only known copy, printed on vellum, with a manuscript entry of sale to Henry Langley and his wife Katherine[3] on 13 December 1476,

---

[1] Caxton the printer attended in May 1480 the biennial audit of the churchwardens' accounts of St. Margaret's for the period 1478-80, in which the funeral fees of the deceased William Caxton were included, and again acted as auditor and attestor in 1482 and 1484.

[2] Edward sent Hastings in person with heavy reinforcements to Calais in February 1477, as a precaution against Louis's invasion of neighbouring Artois. Other such letters of protection were then issued as for the service of William Rosse, the victualler and stapler of Calais, whom Caxton must have known well, as they were employed together in the various trade delegations to Bruges in 1472-5. It is significant of Caxton's high contacts and one-upmanship that he was able to go over Rosse's head to the great Hastings himself.

[3] Henry Langley, of Rickling Hall, near Thaxted, Essex, died on 2 April 1488, and his widow Katherine on 18 October 1511.

was discovered at the Public Record Office in 1928. Indulgence printing had begun with Gutenberg himself in 1454, and was particularly favoured by printers who, like Caxton, were in the throes of installing a new establishment and required jobbing work.

The Pope's commissary for sale of this *Indulgence* from 24 May 1476 (aided by the papal collector of revenues in England, Johannes de Gigliis) was John Sant, Abbot of the great Benedictine monastery at Abingdon from 1468 to 1496, and a good friend of Eastney. Sant had coaxed this valuable sideline from Sixtus IV during a diplomatic mission to Italy in 1474, and also persuaded the Pope to commute Eastney's obligatory visit to Rome for confirmation of his appointment as Abbot, with confiscation of funds meanwhile, into an annual payment of 100 florins.[1] Eastney himself must have put Sant in touch with Caxton, and his revolutionary method of producing Indulgences cheaply in vast numbers within a few hours, instead of hiring an expensive scribe to write them slowly one at a time. But Rivers also was probably concerned, for he had just returned from his own pilgrimage tour of Italy,[2] and Sixtus had appointed him (as Caxton himself stresses in his epilogue to the 1479 *Cordiale*) 'defender and director of the causes apostolic for our holy father the Pope in this realm of England'. Abbot Sant's Indulgence, as usual (if one reads carefully through the preliminary sales talk), grants little more than leave to choose one's own confessor, and does not include sacramental remission of sin. Caxton printed it in his type 2, with the first word (the Abbot's name [I]Ohannes) in the bold new type 3 for display.

Perhaps the Sant *Indulgence* was not the only work produced by Caxton before the close of 1476. The London printer Robert Copland, who worked for Wynkyn de Worde from 1508 and was in a position to know (he may even have been one of Caxton's own employees in early youth),[3] remarked in the preface to his *Apollonius of Tyre* (1510) that he had translated the work from the French 'at the exhortation of my foresaid master [Wynkyn], according directly to mine author, gladly following the trace of my master Caxton, beginning with small stories and pamphlets and so to other'.[4] In fact

---

[1] Sixtus finally ratified the agreement in 1478, after repeated angry letters on Eastney's behalf from Edward.

[2] On 6-8 June 1476 Rivers visited Charles at Morat, but very sensibly left before the battle. He was home in England in time to attend the re-interment of Edward's father Richard and little brother Rutland (slain at Wakefield and beheaded after death in 1460) at Fotheringay on 29 July.

[3] Duff thinks not, because Copland lived till 1548. Still, in that year he was referred to as 'old Robert Copland the eldest printer of England'. His words imply that he wrote from experience, and that Caxton had been his master in the same sense as Wynkyn.

[4] 'So to other' no doubt refers to *Apollonius*, which was Copland's first substantial work. There is a slight confusion of ideas, for none of Caxton's early small pieces are his own translations, but Copland's meaning is clear enough, that he (as a translator) and Caxton (as a printer in England) both began with small pieces before turning to larger works.

IIa Earl Rivers presents his Dicts of the Philosophers to Edward IV.

Jf it plefe ony man fpirituel or temprel to bye ony
ppes of two and thre comemoraciõs of falifburi vfe
enprpntid after the forme of this prefét lettre whiche
ben wel and truly correct, late hym come to weftmo;
nefter in to the almonefrpe at the reed pale and he fhal
haue them good chepe ∴

Supplico ftet cedula

Ib Caxton's Type 3, from his Advertisement for the Sarum Ordinal, 1479.

Thus ende I this book whyche I haue translated after myn Auctor as nyghe as god hath gyuen me connyng to whom be gyuen the laude ⁊
preysyng / And for as moche as in the wrytyng of the
same my penne is worn / myn hande wery ⁊ not stedfast
myn eyen dimed with ouermoche lokyng on the whit
paper / and my corage not so prone and redy to laboure
as hit hath ben / and that age crepeth on me dayly and
febleth all the bodye / and also be cause I haue promysid
to dyuerce gentilmen and to my frendes to adresse to hem
as hastely as I myght this sayd book / Therfore I haue
practysed ⁊ lerned at my grete charge and dispense to
ordeyne this said book in prynte after the maner ⁊ forme
as ye may here see / and is not wreton with penne and
ynke as other bokes ben / to thende that euery man may
haue them attones / ffor all the bookes of this storye named the recule of the historyes of troyes thus enpryntid
as ye here see were begonne in oon day / and also fynyss-
shid in oon day / whiche book I haue presented to my
sayd redoubtid lady as a fore is sayd . And she hath
well acceptid hit / and largely rewarded me / wherfore
I beseche almyghty god to rewarde her euerlastyng blysse
after this lyf . Prayng her said grace and all them that
shall rede this book not to desdaigne the symple and rude
werke . nether to replye agaynst the sayyng of the ma-
tres towchyd in this book / thauwh hyt acorde not vnto the translacōn of other whiche haue wreton hit / ffor
dyuerce men haue made dyuerce bookes / whiche in all
poyntes acorde not as Dictes . Dares . and Homerus
ffor dictes ⁊ homerus as grekes sayn and wryten fauo-
rably for the grekes / and grue to them more worship

a dozen or so small undated quarto works belong to the earliest years of Caxton's Westminster press. These include two editions each of Lydgate's poems *The Horse, Sheep, and Goose*, and *The Churl and the Bird*, the first editions of which must have been among the earliest of all, in order to allow time for them to sell out and for second editions to be required. Stevenson saw that the supposed first editions have the same paper (with pregnant unicorn mark, probably from Saint-Cloud near Paris) as the first few quires of the first edition of *Canterbury Tales*, the remainder of which cannot have been completed before 1478; so he inferred that all this matter was probably printed before the end of 1476, and that *Canterbury Tales* then had to be laid aside, so that Caxton could produce the two commissioned works for patrons, *Jason* and *Dicts*, which occupied most of the year 1477.

Caxton's first large work at Westminster was Raoul Lefèvre's *History of Jason* in his own translation, which (as already suggested) he had perhaps begun at Bruges before or after printing the French original there with Mansion's help.[1] *Jason*, which Lefèvre presented to Philip the Good in the early 1460's, a year or two before *Recueil*, exploits the same convention of rewriting classical myth in the manner of the contemporary romance of chivalry; so that Jason's first exploit is to joust with Hercules at the court of Thebes, 'and never bare that one that other to the earth, whereof all they that saw it had great marvel, forasmuch as Hercules was more in members and higher than Jason was, but Jason was so well on horseback that no man might unhorse him'. The intention was to glorify the Order of the Golden Fleece ('a kind of Burgundian Garter', as Professor Vaughan says), which Philip had founded long ago at his wedding to Isabella of Portugal in 1430. All Burgundy was tapestried and frescoed with the story of Jason, including a room in Philip's favourite palace at Hesdin in Artois,[2] 'wherein was craftily and curiously depainted the conquest of the Golden Fleece,' wrote Caxton in his prologue, recalling old memories, 'in which chamber I have been, and he had do make in the said chamber subtle engine that when he would it should seem that it lightened and then thunder, snow and rain'. Philip's practical jokes at Hesdin also included devices for showering the visitor with water, flour and soot, mechanical figures to beat him with sticks, a bridge that hurled him into the water, a wooden talking hermit to exhort him to virtue, and squirts in the floor 'to wet the ladies from underneath'. Such bonhomies perhaps gave Philip his title of the Good, which he scarcely earned for any other reason.

[1] No one has yet bothered to edit Caxton's French editions of *Recueil* and *Jason*, and to discover whether he used the same manuscripts for his translations, or whether he translated *Jason* from the printed edition.

[2] Hesdin is on the river Canche, southeast of Boulogne, halfway between Crécy and Agincourt. A rival, antipagan view, that the Fleece was Gideon's (see *Judges* 6.36-40), was favoured by two bishop chancellors of the Order, Jean Germain and Guillaume Fillastre, 'wherein,' says Caxton prudently, 'I will not dispute.'

III Caxton's Type 1 from Recuyell of the Histories of Troy, 1475.

As Edward himself was a knight of the Golden Fleece, 'whereof our said sovereign lord is one and hath taken the profession thereof', no literary dish could be better chosen to set before the King, and to link Caxton's past in Bruges with his future at Westminster. After namedropping Duchess Margaret and her patronage of *Recuyell*, Caxton proceeded: 'then for the honour and worship of our said most redoubted liege lord which hath taken the said Order I have under the shadow of his noble protection enterprised to accomplish this said little book, not presuming to present it to his Highness, forasmuch as I doubt not his good grace hath it in French which he well understandeth, but not displeasing his most noble grace I intend by his licence and congee and by the supportation of our most redoubted liege lady the Queen to present this said book unto the most fair and my most redoubted young lord my lord Prince of Wales, our tocoming sovereign lord, to the intent that he may begin to learn read English'. It was no doubt Anthony Rivers who had procured this permission from his sister the Queen and his brother-in-law the King to dedicate *Jason* to his little ward the Prince, born in the sanctuary of that same Abbey in the troubles of 1470,[1] and now aged six. Edward had appointed Rivers governor to the Prince in November 1473, with orders that he should be instructed from breakfast to dinner in 'such virtuous learning as his age shall suffer to receive', and during dinner listen to the reading of 'such noble stories as behoveth a prince to understand and know'; so *Jason*, loaded with sex and violence though it was, would come in very useful. In his epilogue Caxton with characteristic editorial zeal added further information on Jason ('this have I founden more than mine author rehearseth in his book') from Boccaccio's *Genealogia deorum* (which his master Veldener had printed in 1473)[2]. He ended with a prayer for the little Prince, 'whom I beseech God Almighty to save and increase in virtue now in his tender youth that he may come unto his perfect age'; but this last request was not to be granted.

The patron and translator of Caxton's next book, *Dicts of the Philosophers*, 18 November 1477, was none other than Rivers himself. Something had come over the romantic Earl in 1471, after his wound at Barnet and his forlorn hope sortie that saved London from Fauconberg. Perhaps it was the shock of Warwick's chopping his father's and brother's heads off, perhaps it was battle-weariness, but probably it was most of all the humiliation of being sacked by Edward that July from his governorship of Calais, and being replaced

[1] Eastney, who acted as godfather with Abbot Milling at the Prince's christening, remained a special friend of the grateful Queen. She endowed the new chapel of Saint Erasmus (patron saint of childbirth, because he was martyred by having his intestines wound out on a windlass) at the Abbey in 1479, and Eastney gave her sanctuary again when Gloucester seized power and murdered her sons in 1483.

[2] However, although he could cope quite well with Latin, Caxton perhaps read this work in Laurent de Premierfait's French translation, which was not printed until 1499, but was available in manuscripts.

by his enemy Hastings, and Edward's calling him a coward when he put in for leave to go to Portugal and fight Saracens. Although Edward still humoured him with showy jobs and minor missions, Rivers now turned to pilgrimage and authorship. In 1473, 'subject and thrall unto the storms of fortune, and perplexed with worldly adversities of which I Anthony Woodville have largely and in many different manners had my part', he made the golden journey to St. James at Compostella, and remembered it in words which show that he too—like Caxton, and many other medieval writers, but so rarely—could regain Time Lost. 'I shipped from Southampton in the month of July the said year, and so sailed from thence till I came into the Spanish sea, there lacking sight of all lands, the wind being good and the weather fair, then for a recreation and a passing of time I had delight and asked to read some good history. And among other there was in my company a worshipful gentleman named Louis de Bretailles,[1] which greatly delighted him in all virtuous and honest things, that said to me he hath there a book that he trusted I should like it right well, and brought it to me, which book I had never seen before, and is called *The Sayings or Dicts of the Philosophers.*'

This *Dits des philosophes*, another Burgundian favourite which Mansion himself printed at Bruges soon after Caxton's English edition, had a complicated ancestry. The original was an Arabic compilation of mostly apocryphal and unimaginatively feeble sayings of ancient sages (some real, like Solon, Pythagoras, Hippocrates, Socrates, Plato, and some nonexistent, like 'Sedechias', 'Tac', 'Salquinus'), written about 1050 by a certain Mubashshir ibn Fatik of Damascus and called *Mukhtar al-hikam*; this was translated into Spanish as *Bocados de oro* (Golden Mouthfuls), thence into Latin about 1250 as *Liber moralium philosophorum* by Emperor Frederick II's physician Johannes de Procida, and at last about 1400 by Guillaume de Tignonville, Provost of Paris, into French, in the *Dits des philosophes* given to Rivers by his Gascon friend.

'After such season as it listed the King's grace command me to give mine attendance upon my lord the Prince,' Rivers continues, 'when I had leisure I looked upon the said book, and at the last concluded in myself to translate it into the English tongue, which in my judgment was not before,[2] thinking also full necessary to my said lord the understanding thereof.' So *Jason* and *Dicts* belong together as joint attempts by Caxton and Rivers to provide printed

[1] A Gascon knight who had fought on the English side in the Hundred Years' War, took part in River's Anglo-Burgundian tournament at Smithfield in 1467, and served Edward as a diplomatic agent throughout the reign.

[2] Two other translations existed of which Caxton and Rivers were unaware, including Stephen Scrope's made in 1450 for the Paston family's patron Sir John Fastolf. Blades and others are mistaken in thinking they plagiarised Scrope's version and pretended they had never heard of it. Comparison shows that apparent resemblances are coincidental, being due to the prevalent habit of taking over any difficult French words into English, and that Scrope's many errors are not repeated by Rivers.

reading matter for the heir to the throne—or rather, for the wider public who would be eager to read the same books as the little Prince.

Caxton takes up the story in his epilogue. Rivers sent him his translation 'in certain quires' (that is, in unbound manuscript or fair copy) 'to oversee'. Caxton found it 'according unto the books made in French which I had oft afore read, but certainly I had seen none in English till that time'. He called on Rivers, who despite his polite protests ('I answered unto his lordship that I could not amend it, but if I should so presume I might impair it, for it was right well and cunningly made and translated into right good and fair English') 'willed me to oversee it and also desired me that done to put the said book in emprint. And thus obeying his request and commandment I have put me in devoir to oversee his said book and behold as nigh as I could how it accordeth with the original being in French.'

Much to his amazement and amusement Caxton discovered that Rivers had omitted the aspersions of Socrates against women. How could this be? 'I suppose that some fair lady hath desired him to leave it out of his book, or else he was amorous on some noble lady for whose love he would not set it in his book, or else for the very affection, love and good will that he hath unto all ladies and gentle-women, he thought that Socrates spared the sooth and wrote of women more than the truth.' In any case, Caxton observed with tongue in cheek, Socrates's strictures were hardly applicable to Englishwomen, 'for I wot well of whatsomever condition women be in Greece, the women of this country be right good, wise, pleasant, humble, discreet, sober, chaste, obedient to their husbands, true, secret, steadfast, ever busy and never idle, temperate in speaking, and virtuous in all their works, or at least should be so.' On reflection ('I am not in certain whether it was in my lord's copy or no, or else peradventure that the wind had blow over the leaf at the time of translation'), and with a plea to his readers 'if they find any fault to arrest it to Socrates and not to me', Caxton translated the offending matter[1] himself at the end of his epilogue, and concluded with an acknowledgment of 'the good reward that I have received of his said lordship'.

Caxton's elaborate Chaucerian pleasantry about Rivers and women was one of the best jokes he ever made,[2] so more is the pity that its real topical point has not been appreciated since his time. It is inconceivable that Caxton would have dared to involve his patron in the jest without first obtaining his full permission, or that Rivers would have joined in unless he felt it showed him in a usefully

[1] It is rather an anticlimax. The real Socrates could have done better than 'Women be the snares to catch men, but they take none but them that will be poor or else them that know them not', and so on. Rivers himself points out his omission by remarking (leaf 32 recto: 'and the said Socrates had many sayings against women which is not translated'.

[2] 'Or at least should be so' makes me laugh every time, and was fit to drive Caxton's wife Maud to burn her wimple.

flattering light. This it did. Rivers's first wife, Lady Scales,[1] had died in September 1473 while he was away at Compostella, and left him for the next ten years the most eligible widower in England. His greatest opportunity had arrived and vanished just before Caxton printed D*icts*. Charles the Bold was killed besieging Nancy in Lorraine on 5 January 1477,[2] Louis immediately invaded his dominions, beginning with French Burgundy and Artois, and it became urgent to find a husband to defend Mary of Burgundy, Charles's only child (by his second wife, Isabel de Bourbon) and his successor. The Dowager Duchess Margaret of course suggested fleeting Clarence, who himself had just become a widower in suspicious circumstances. Edward vetoed this politically impossible proposal, and sent their mutual friend Louis de Bretailles to Ghent in May to offer the hand of Rivers complete with an army. But Mary, who was then twenty and legally old enough to choose for herself, very sensibly rejected him and married Maximilian the Emperor's son on 19 August. The *Dicts* byplay is meant to show that now, so far from his nose being out of joint, Rivers is still the perfect knight and lady's man, and still in the matrimonial market. Sure enough in December 1478 King James III of Scotland sent an embassy to Edward proposing that Rivers should marry his flighty sister Margaret; but the two monarchs kept Rivers dangling for four years, and in the end turned him down, so that he had to make do with a mere heiress.

Towards the end of the printing of the last page of *Dicts* Caxton added a colophon (surviving only in the John Rylands Library copy at Manchester), reaffirming that Rivers had commanded him to have the work 'set in form [i.e. type] and emprinted in this manner as ye may here in this book see', and dated 18 November 1477. This stop-press variant must have been printed off within a few hours of the rest, and the date therefore holds good for the completion of the whole edition.[3] Meanwhile a scribe named Haywarde was writing the royal presentation manuscript, now in Lambeth Palace Library, copying the main text not from the printed edition but from another manuscript. At the end Haywarde wrote, evidently in collusion with Rivers and Caxton: 'And one William Caxton at desire of my Lord Rivers emprinted many books after the tenor and form of this book, which William said as followeth . . .', and added Caxton's epilogue from the printed edition, complete with Socrates on women. Haywarde's illuminated frontispiece[4] shows Rivers himself on bended

[1] Anne, daughter of the Lancastrian Lord Thomas Scales who kept the Tower for Margaret in 1460 and was murdered by the mob. Rivers married her soon after, before he turned Yorkist, and took her father's title.

[2] His Portuguese physician and his page found his body several days later, stripped naked by looters and with its face gnawed by wolves.

[3] So it is a bibliographical solecism to call the colophon variant a re-issue, as Crotch and others do.

[4] The young man in royal ermine in the background has not hitherto been identified, but must be Gloucester, as Clarence was then already in disgrace and imprisoned in the Tower.

knee in an uncomfortable suit of plate armour, with the tonsured priest Haywarde on both knees behind him, handing the book to Edward on his throne with crown, orb and sceptre, and the Queen and little Prince at his side. Haywarde finished his task on 24 December, just in time for a Christmas present.

Clarence had taken his rejection far harder than Rivers; indeed, his subsequent conduct seems inexplicable unless by disappointed rage at having murdered his wife all for nothing in expectation of a Burgundian marriage. Duchess Isabel, Warwick's daughter, had died very conveniently on 22 December 1476, two months after a difficult childbirth, but just in time to free Clarence for eligibility.[1] In April 1477 Clarence had her lady attendant Ankarette Twynho kidnapped in Somerset, carried off to his home base at Warwick Castle, charged before an intimidated jury with poisoning her, and hanged. In May, while Edward was counter-hanging his wizard henchmen Stacy and Burdett, Clarence sponsored an unsuccessful rising in Cambridgeshire led by a bogus Earl of Oxford (the real one, Caxton's future patron, was still safe behind bars at Hammes near Calais). Clarence's old dream of carving up Burgundy with Louis and deposing Edward had revived and failed; Louis, instead of cooperating, warned Edward in June of his treachery, and even alleged that Duchess Margaret was in league as ever with their false brother. Edward, always the most ruthless operator of them all when driven to it, clapped Clarence in the Tower in June 1477, and appeared in person to attaint him of high treason before a special Parliament in January 1478. At that moment Rivers himself was on top of the world, as best man at the four-year-old Duke of York's wedding to the five-year-old heiress Anne Mowbray at St. Stephen's Chapel Westminster on 15 January, and as champion at the great tournament in Palace Yard on the 22nd, where he challenged all comers disguised as the White Hermit in a portable black velvet hermitage, and subsidised gentlemen who complained that the entrance fee was too high. No doubt Caxton was there in the grandstand. On 18 February 1478 Clarence was duly executed, quite possibly by drowning at his own choice, kinky to the end, in the celebrated butt of malmsey. It was surely Rivers himself who chose the moment, only two days later, for Caxton to produce a booklet calculated to show the moral superiority of the King's brother-in-law over the King's late brother.

On 20 February 1478 Caxton completed the printing of another translation by Rivers, the *Moral Proverbs* of the amiable French literary lady Christine de Pisan (1364-1430), a short work of only four leaves in decasyllabic couplets, which can only have taken two or three days to print. Caxton added a colophon of his own in rather engaging rhyme royal doggerel:

[1] Duchess Isabel had been expendable ever since 1474, when Parliament gave Clarence possession of her estates in the event of her dying before him, and still more so after producing his son and heir Edward, Earl of Warwick, on 21 February 1475.

Go thou little quire and recommend me
Unto the good grace of my special lord
The Earl Rivers, for I have emprinted thee
At his commandment, following every word
His copy, as his secretary can record,
At Westminster of Feverer the xx day
And of King Edward the xvii year vraye.
                    Emprinted by Caxton
                    In Feverer the cold season.

As for the moral Christine, *'An happy house is where dwelleth prudence,'* she unexceptionably thinks, and:

He that seeketh often other to blame
Giveth right cause to hear of him the same.

Much of the year 1478 must have been taken up with the completion of the enormous *Canterbury Tales* which Caxton had perhaps already begun and laid aside in 1476. As a first printing of a national poem in the fifteenth century this is comparable only to the first edition of the *Divina Commedia* (1472) or Villon (1489), and for the monumental magnificence of the type page combined with the ever-new impact and wonder of the text this is perhaps Caxton's finest book (though *Morte d'Arthur*, 1485, is a good second). In the prologue to his second edition, in 1483, Caxton wrote an apology for his first: 'I find many of the said books (i.e. manuscripts of *Canterbury Tales*) which writers have abridged it and many things left out, and in some place have set certain verses that he never made nor set in his book, of which books so incorrect was one brought to me six year past, which I supposed had been very true and correct, and according to the same I did do emprint a certain number of them, which anon were sold to many and diverse gentlemen.' As a succinct and accurate view of the problems of Chaucerian textual criticism Caxton's opening words are admirable and unique in his time, although modern scholars dispute his editorial knowledge and concern as though he were taking the bread out of their mouths. Still, in the intention of advertising his second edition with this highminded blurb, he exaggerated a little. In fact Caxton's first manuscript was a rather inferior member of Group 'b' in the present classification according to the variable order of the tales, just as his second was an inferior member of Group 'a', and each was a reading text of normal quality in that time, with an ordinary quota of omissions and interpolations.

Perhaps (as Blake suggests) Caxton had a patron for *Canterbury Tales*, for in *Morte d'Arthur* and other works he uses a similar expression, that the manuscript was 'unto me delivered', apparently to mean that it was supplied by a patron. But if it was merely lent to him by a friend or by the owner of a scriptorium, would he have put it differently? Caxton's persistence in printing Chaucer's works suggests a personal venture, motivated both by ready profit and by admiration, though it would not be surprising if for the sale and distribution of these and other popular, non-courtly pieces he had secured the support of a bookseller in the manuscript trade.

Another Chaucer work immediately followed. Boethius, *Consolation of Philosophy*, in Chaucer's English translation, perhaps prompted by Mansion's French edition of 28 June 1477, must have appeared towards the end of 1478, since it is Caxton's last large book in the first state of his type 2, which was superseded by the second state (type 2*) early in 1479. The gothic type 3 is here used for the first time with its intended function as heading and Latin type; perhaps he had not possessed a complete stock of it when he arrived in 1476 and included only six letters of it in the Sant *Indulgence*. Caxton printed *Boethius*, as he says in his epilogue, 'at request of a singular friend and gossip of mine', who may well have been his livery companion of 1452, William Pratt, mercer, whom he mentions in similar terms in *Book of Good Manners*, 11 May 1487. Perhaps Pratt, if it was he, took a financial stake in the edition; perhaps he also supplied the manuscript, as he did that of *Book of Good Manners*; or perhaps this was merely Caxton's way of dedicating the book to a likeminded friend. By a happy coincidence Chaucer's tomb was in the Abbey only a few yards from Caxton's shop, being just round the corner in front of St. Benedict's chapel, the first to the right on entering the south door of Poets' Corner.[1] Caxton, with feelings of true reverence indistinguishably combined with his instinct for publicity, commissioned a Latin epitaph for the poet, ending with a quatrain in praise of his own Chaucer editions,[2] from an itinerant Italian humanist and poet laureate from Milan, Stephanus Surigonus, and had this inscribed on a tablet and hung on a pillar next to the tomb, and also printed it in his epilogue.[3] This Surigonus was used to such work for printers, having written a Latin verse advertisement for the first edition of Virgil printed by Johann Mentelin at Strassburg about 1469.[4] Surigonus had matriculated at the University of Cologne in

[1] Chaucer's tomb was moved a short distance in 1556 to its present location, and became the nucleus of Poets' Corner itself through the practice of burying poets, beginning with Spenser in 1599, as near as possible to Chaucer.

[2] 'Caxton wished you to live after death, illustrious poet Chaucer, through the care of your own William, for he not only printed your works in type, but also ordered this praise of you to be placed here' (*has quoque sed laudes jussit hic esse tuas*).

[3] Caxton states clearly that the Latin verses are copied from the tablet ('by whose sepulture is written on a tablet hanging on a pillar his epitaph made by a poet laureate whereof the copy followeth') and the verses themselves affirm that they were put there by his order. Other notions (that Surigonus wrote them before 1470 independently of Caxton, that the quatrain about Caxton does not belong to the tablet verses, that it was written by Caxton himself, that he did not have it included on the tablet but only inserted it in the printed *Boethius*) seem groundless and gratuitous.

[4] Included in a collection of Surigonus's tedious verses in British Library MS. Arundel 249, ff. 94–117. Mentelin's *Virgil* disputes its claim for priority with Sweynheym and Pannartz's at Rome in the same year.

1471, after an earlier visit to England including Oxford in the period 1454-64; he hung around the Low Countries and the court of Charles the Bold in the early 1470s, and had recently revisited England, teaching at Cambridge in 1475-6; so Caxton may well have met him in any or several of these avatars.

Meanwhile, before the end of 1478, because they are all still in the first state of type 2, Caxton had produced a dozen small undated quartos, including two minor works of Chaucer, *Queen Anelida and the False Arcite*, and *Parliament of Fowls*, second editions of Lydgate's *Churl and Bird* and *Horse, Sheep and Goose*, and—as further instalments in his project to add his second favourite poet to his Chaucer series ('I that am not worthy to bear his [Lydgate's] penner and ink horn after him', he had written in *Recuyell*)—Lydgate's *Temple of Glass* and *Stans puer ad mensam*. *Stans puer* is Lydgate's translation of a standard Latin manual of table manners for boys by Robert Grosseteste, Bishop of Lincoln in 1235-53, which Blades and others have confused with a completely different text, the *De moribus puerorum* by the fifteenth-century Italian humanist Johannes Sulpitius Verulanus.[1] As a companion piece Caxton produced a verse *Book of Courtesy* by an anonymous follower of Lydgate, also called *Little John* after the little victim to whom it is addressed ('*Little John, since thy tender infancy* . . .'). Both these were established first English reading books for schoolchildren, and were perhaps commissioned by a bookselling schoolmaster, possibly Otuel Fuller, a married layman who in 1474-82 was master of the school in the Almonry which flourishes to this day as Westminster School.[2] A schoolbook of international popularity was Cato, *Disticha*, a collection of moral couplets in respectable late classical Latin hexameters, to which the name of a nonexistent Dionysius Cato became attached in allusion to the ancient Roman family proverbial for wisdom and virtue. Caxton printed two quarto editions in 1477-8, and a third in folio towards 1482, of an English verse translation, complete with the Latin original, made by Benedict Burgh. Caxton probably knew the translator, and perhaps printed the book to his order. Benedict Burgh, a friend and collaborator of Lydgate in youth, whom we have already met in the Caxton deeds of Little Wratting, together with the very patron for whom this translation was written about 1440,[3]

---

[1] *Stans puer* (for the Latin text see F. J. Furnivall, *The Babees Book*, 1868, II, 30-33) begins: *Stans puer ad mensam domini bona dogmata discas*, and ends in some manuscripts (e.g. British Library MS. Add. 37075) with the attribution to Grosseteste: *Haec qui me docuit grossum caput est sibi nomen.* Sulpitius's poem begins: *Quos decet in mensa mores servare docemus.*

[2] E. J. L. Scott, favoured by Plomer, Crotch, and others, produced the unfounded notion that Fuller lived in the house called Saint Albans near Caxton's Chapter House shop, and identified him with the Schoolmaster Printer of St. Albans, Hertfordshire, with whom Caxton was associated.

[3] Burgh appears in a Caxton deed of 1456 together with Henry Bourchier later Earl of Essex (1404-83), a Yorkist magnate and cloth-trader. Essex's son and heir Sir William Bourchier, for whose tuition Burgh wrote his *Cato*, married Rivers's sister Anne Woodville about 1467, and died in 1483.

became a canon from June 1476 until his death on 13 July 1483 at St. Stephen's chapel in the Palace of Westminster just over the way from Caxton's shop. The Latin text is normally divided, as in the Burgh-Caxton version, into a short introduction called '*Parvus Cato*', and the main work, '*Magnus Cato*'. Both parts occur in the oldest surviving manuscripts dating from the Carolingian period, towards A.D. 800. The notion that '*Parvus Cato*' was added about 1180 by Daniel Church, a priest at the court of Henry II, is a mere eighteenth-century absurdity, concocted by Thomas Warton, repeated by Ames-Herbert-Dibdin-Blades-Duff, and never yet corrected.[1]

Two more of these early quartos in type 2, *Infantia Salvatoris* (a legend of the childhood of Christ from the New Testament Apocrypha) and *Propositio Johannis Russell*, are in Latin; perhaps this is another indication that type 3 was not yet available in quantity in 1476-7. The *Propositio* is the speech made by Caxton's diplomatic colleague John Russell for the investiture of Charles the Bold with the Garter at Ghent on 4 February 1470, and presents an unsolved bibliographical problem. Where and when was it printed? Blades thought, and no one has liked to contradict him outright, that it 'was issued at Bruges at no long period after its delivery', because 'at that time, both with the subjects of the Duke of Burgundy and the "English Nation" there resident, it would secure a good circulation: not so if issued seven years after its delivery in another country'. In fact the *Propositio* could not have been produced at Bruges until the first appearance of type 2 in the French *Cordiale* towards the early summer of 1476, and could have been printed at Westminster from Michaelmas 1476 onwards; so the time-gap is not seven years, but three months! Nor would Russell have expected a rousing reception from the subjects of Charles the Bold, who had cut Edward since the French invasion fiasco of 1475, and bitten his Garter to pieces in fury.[2] The *Propositio* was not intended anyway for the general reading public; it is a rare English example of a class of tracts of which many hundreds were produced abroad, especially at Rome, the Latin oration privately printed for its speaker to distribute among his friends and prospective benefactors. If Caxton *did* print it at Bruges soon before his departure, as is just possible, then it must have been at Russell's order for export to England; he had no doubt met Russell repeatedly at Bruges during his many diplomatic visits, from the marriage feelers in April 1467 (when Russell bought the Mainz Cicero, *De officiis*, at a bookshop in St. Donatian's)

---

[1] Church wrote a totally different work called *Magnus Urbanus*, to which somebody added a *Parvus Urbanus*. Warton mixed these up with *Cato* in his *History of English Poetry* (1774-81), and everyone copied him.

[2] Edward, by way of tit for tat, stopped wearing his Golden Fleece, until Maximilian sent a special envoy in 1478 to ask him why. Did Caxton miscalculate when he printed the Fleece-promoting *Jason* in 1477? Perhaps Duchess Margaret put him up to it in aid of Anglo-Burgundian rapprochement, and perhaps Rivers hoped for a renewal of the £1000 pension which the late Charles had given him in 1471.

to the Hanseatic treaty in 1473-4, when Caxton was already beginning to instal his first press. But a Westminster origin, probably early in 1477, seems much more likely. Russell had just been made Bishop of Rochester, on 6 September 1476, and was on his way up again. Burgundy was in the news once more, Charles was mentionable, being dead, Rivers was the official English candidate for his daughter's hand. Perhaps Rivers had a hand in the publication, for Russell was his old colleague in the Bruges missions of 1467-8, and was appointed in December 1478 to negotiate his marriage contract with Margaret of Scotland. The *Propositio* has the same type-page measurements as the other early Westminster quartos, and therefore probably came off the same press at the same time; and one of the two surviving copies (in the John Rylands Library, Manchester) was bound in a volume of English manuscript pieces written by the Westminster scribe William Ebesham.[1]

Caxton's dozen early quartos total only about 230 leaves, and some or most could easily be fitted (as Copland's remark implies, 'beginning with small stories and pamphlets and so to other') into the period before *Jason* and *Dicts*, from autumn 1476 to spring 1477— even though, as nearly all survive in unique copies and none in more than two copies, it is statistically likely that one or two more editions existed and are now lost. I have noticed that in *Parliament of Fowls* the compositor ran short of capital T, and was forced to borrow from type 3;[2] and as the same mixture still persists in *Anelida and Arcite* and *Book of Courtesy*, these three were probably printed together. A broken final d is often found from *Jason* onwards, but the letter seems undamaged in the quartos, excepting *Temple of Glass*.

In February 1479 Caxton discarded the first state of his type 2 for a new casting, type 2*, evidently made by Veldener from the same matrices or punches as before, but trimmed to remove various imperfections and to produce a slenderer outline, and augmented with a number of new forms.[3] Blades had the preposterous notion that Caxton cast the new state himself, using the old type as punches to strike the matrices. Perhaps Caxton saved type 2* in honour of his patron for the first dated book in which it appears, which is a *Cordiale* in English translation by Rivers. In his epilogue Caxton mentions with justified pride that Rivers delivered the manuscript to him on 2 February 1479; he began printing next day, and finished on 24 March, a creditable total of seventy-eight leaves in fifty-one days. This was the time when Rivers seemed as good as betrothed to

[1] And therefore not, as has been suggested, in a Netherlandish hand, which if true would have favoured Bruges.
[2] The same accident occurs with S in *Dicts*, and with H in *Moral Proverbs* and *Boethius*.
[3] The easiest dodge for distinguishing the two states is to look for the letters th, which are printed from two separate sorts in the first state, but from a single ligatured sort in the second. A defective A with three prongs instead of two at bottom left is frequent in state one, but very rare in 2*. A new alternative k with head hooked instead of looped appears in 2*.

Margaret of Scotland, and Caxton's renewed propaganda for his virtues and capacities was no doubt concerted with his patron, particularly the repeated mysterious allusion to 'the time of the great tribulation and adversity of my said Lord'. Ever since then, 'he hath been full virtuously occupied, as in going of pilgrimages to Saint James in Galicia,[1] to Rome to St. Bartholomew, to St. Andrew, to St. Matthew in the realm of Naples,[2] and to St. Nicholas of Bari in Apulia, and other diverse holy places, and also hath procured and gotten of our Holy Father the Pope a great and a large Indulgence and grace unto the chapel of Our Lady of the Pew by St. Stephen's at Westminster'. Rivers had restored this chapel, destroyed by fire in 1452,[3] and wished to be buried there if he should die south of Trent; but he was not to die south of Trent. Despite these pious travels, and his 'great labours and charges in the service of the King and of my said Lord Prince as well in England as in Wales',[4] Rivers has 'put him in devoir at all times when he might have a leisure, which was but start-meal, to translate diverse books out of French into English. Among other passed through mine hand the book of the wise sayings and dicts of philosophers, and the wise and wholesome proverbs of Christine of Pisa set in metre, over that hath made diverse ballads against the seven deadly sins.'[5] Caxton adds that he has obeyed Rivers's command to print *Cordiale* because 'I am bounden so to do, for the manifold benefits and large rewards of him had and received of me undeserved'.

The *Nova Rhetorica* or *Margarita eloquentiae*, a treatise on the art of Latin speechmaking by the Italian humanist Laurentius Traversagni de Saona, doubtless belongs to the same year 1479, for it is printed in type 2* with the uneven line-endings which Caxton did not abandon until 1480. The author, a Franciscan from Savona near Genoa, had completed his text on 26 July 1478[6] at Cambridge, where he lectured on and off from 1476 to 1482. Perhaps he was introduced to Caxton by Surigonus. Technically this is one of the most important of all Caxton's books, as it is the only one of which we know for certain the manuscript copy-text from which it was actually printed, with the evidence which this reveals concerning the processes of printing.[7]

[1] Compostella, July 1473, when Rivers first read *Dicts*.
[2] St. Andrew at Amalfi, St. Matthew at Salerno. Rivers's Italian tour in 1476 was somewhat marred by an encounter with bandits near Rome, who robbed him of jewels 'worth a thousand mark or better'. Sixtus helpfully announced a pardon for the thieves and a curse on the receivers. The Venetian senate bought in some and restored it free of charge.
[3] A boy from the Almonry school had been detailed to put the candles out on 17 February 1452, but forgot.
[4] The Prince resided at Ludlow Castle on the borders of his Principality with a view to keeping the wild Welsh in order.
[5] Caxton's words rather suggest that he expected to print these ballads, which are now lost. He may well have done so without the edition surviving.
[6] Caxton's '6 July' is a misprint. If you look at the author's manuscript you can see that his compositor mistook the 2 for an ampersand.
[7] A few copy-texts are known which were similarly marked off and used for editions by Wynkyn de Worde and Pynson in the 1490s.

The author's autograph manuscript, which was discovered in the Vatican Library by J. Ruysschaert, has been marked by Caxton's press editor with marginal figures prescribing the future contents of each page in the as yet unprinted quires. It is now gradually becoming understood that this process of 'casting off', which a modern printer still performs on an author's typescript, was already normal and basic in fifteenth-century printing. It was necessary when a text had to be divided up for simultaneous printing on several presses, as was Caxton's own practice with large books from *Recuyell* onwards.[1] The most complex and accurate casting off was required for printing 'by the forme', when, in order to economise type and to enable two or more compositors or even presses to share work on the same quire, each side of the sheet (containing two mostly non-consecutive pages in a folio or half-sheet quarto, four in quarto set by the sheet) was set up in type at one time. Internal evidence of printing by the forme is so frequent in fifteenth-century books as to suggest that this was probably the norm, and that seriatim printing (which is often mistakenly taken for granted by bibliographers and textual editors) was probably much less usual.

[1] The practice must have been familiar even to Gutenberg, when he printed his Bible on at least four presses simultaneously.

11

AT THE RED PALE

IT is curious that (with the double exception of the *Sarum Ordinal* and its *Advertisement*, which we shall grapple with in a moment) there seem to be no Caxton editions after *Cordiale* and *Nova Rhetorica* which can reasonably be assigned to the year 1479. An Exchequer issue of 15 June 1479 records a payment to Caxton of £20 'paid to his own hands for certain causes and matters performed by him for the said Lord the King'. This large sum, as other issues of that period show, was a regular advance payment for a mission abroad. Possibly Caxton was sent that summer to Flanders, where Edward was obliging his sister Margaret by giving secret help against Louis, besides tidying up details in the new commercial treaty removing restrictions against English cloth, which the deceased Charles had denied so long, and Maximilian had agreed in July 1478. Then again, 1479 was a plague year in England, and the disruptive effect of plague on printing in fifteenth-century Europe is well known. Another crucial event for Caxton was the beginning of his enormous *Ovid* translation of 220,000 words, which he completed on 22 April 1480, and must have commenced no later than the summer of 1479. Was this a response to some interruption of his press? He had abandoned translation for three years, ever since *Jason*, and now he was to translate with unremitting labour and gusto till the end of his life. Some radical reorganisation of his press must have been necessary to enable him to do so. Perhaps it was really towards this time that Wynkyn de Worde took over as press-manager, and *not* as early as 1473-4? On 4 November 1480, at a meeting of the Abbey Chapter in the Chapter House just behind Caxton's shop, Abbot Eastney granted to Richard Aleyn rector of Fulham a lease of a tenement in the Sanctuary formerly held in the name of 'Elizabeth, wife of Wynand van Worden'.[1] This is the earliest known document which actually names Wynkyn, and suggests that the *latest* possible date for his settling at Westminster might be about 1478-9.

As Caxton seems first to have acquired a full stock of type 3 for the *Boethius* late in 1478, the *Sarum Ordinal* which is his first book entirely in this type probably belongs to 1479; it cannot be significantly later, for it still has the uneven line-endings which Caxton abandoned in the spring of 1480. At the same time, in the same type by way of specimen, and with uneven line-endings, Caxton printed a brief 7-line *Advertisement* of a kind produced by many fifteenth-century printers for nailing on the doors of churches, bookshops, and travelling salesmen's inns. 'If it please any man spiritual or temporal

[1] Elizabeth died in 1498, and their son (?) Julian de Worde in 1500, according to the burial records of St. Margaret's Westminster.

to buy any Pyes of two or three commemorations of Salisbury use emprinted after the form of the present letter which be well and truly correct, let him come to Westminster into the Almonry at the Red Pale and he shall have them good cheap. *Supplico stet cedula* [please do not remove this handbill].'

The Ordinal or Pye (so called from its pied, black and white appearance, like a magpie's) was a kind of perpetual calendar, giving instructions covering each of the thirty-five possible permutations of the liturgical year on procedure when services for annual saints' days clashed with weekly services. A complete Pye of two or three commemorations dealt not only with the two obligatory weekly commemorations of the Blessed Virgin on Saturday and St. Thomas of Canterbury on Thursday, but also allowed for a third commemoration of a patron saint on Tuesday. The text used in Caxton's edition was compiled (as is shown by a complete manuscript in the British Library, Add. 25456) about 1450 by a certain John Raynton for Thomas Gascoigne, successively Chancellor of York Cathedral (1432-42) and of Oxford University (1442-5) and was already long out of date, for it had been superseded about 1455 by Clement Maydestone's *Directorium sacerdotum*. Caxton no doubt produced his Pye to order from some old copy which had given satisfaction for a generation with no complaints from anyone; but he printed Mayde-stone's revised text twice when he got round to it a few years later. Caxton's *Ordinal* narrowly escaped total destruction, as did so many liturgical books from Caxton and other printers when Henry VIII's Reformation made them both useless and illegal. Only a mutilated 8-leaf fragment survives, discovered in 1858 by Blades in the damp-mouldered binding of a Caxton *Boethius* at St. Albans Grammar School.[1]

The *Advertisement* shows Caxton in possession of his famous sign of the Red Pale in the Almonry. Yet is is not until 1482-3 that the Abbey accounts show him in occupation of a room above the Almonry gate rented out by the Almoner, together with an adjacent house rented out by the Prior. Some have been inclined to date the *Ordinal* and *Advertisement* to 1482 or 1483 from the supposed beginning of Caxton's tenancy in the Almonry. To my mind it is the other way about: the telltale line-endings show that these pieces were printed before 1480, the use of type 3 suggests 1479, and we have to date Caxton's tenancy from the *Advertisement* as not later than 1479.

A closer look at the documents shows that this is quite feasible. The Prior in 1482 was a new man, Robert Essex, who made various reforms. Essex reorganised the school in the Almonry (which was probably one of the outlets for Caxton's schoolbooks), keeping the choir boys there, but moving the grammar scholars to a house in the

---

[1] This press-waste, made up of fragments from thirteen Caxtons of the period 1477-83, including two unique, was acquired by the British Museum in 1871. Blades gave the empty binding to St. Bride's Library. Comparison of the *Ordinal* fragment with the complete text in MS. Add. 25456 shows that a perfect copy of Caxton's quarto edition would contain about 130 leaves.

present Dean's Yard, where Westminster School is centred to this day; he took over the Almonry gate loft in 1482 from the Almoner (who continued until 1487 to give it a nil return in his accounts, 'because it is now in the Prior's hands'); and he kept regular entries in his rent book, which had been left empty since Prior William Walsh's days in 1455-6, and in particular throughout the priorate of his immediate predecessor Thomas Arundel in 1474-82. Caxton's Almonry loft and house had been taken on a forty-year lease in 1446 by David Selley, vintner and citizen of London.[1] Selley died soon before 1474, when the will of his widow Cecily was proved, and the lease was held thereafter by his executors, including Walter Lokyngton, who became a monk at the Abbey at the same time as Richard Caxton, in 1473-4. Lokyngton paid the rent in 1479-80. There is no reason why Caxton should not have sublet from Lokyngton; for if he did, his name would not be mentioned in the Almoner's accounts for the loft, because Lokyngton paid, still less in the rent to the Prior for the house, not only because Lokyngton paid for this also, but because Prior Arundel did not bother to keep accounts.[2] From the evidence of the *Advertisement* it seems certain that Caxton did sublet from Lokyngton no later than 1479, and thereafter took over the unexpired seven years of Selley's lease in his own name. Indeed, his subletting may have begun still earlier, perhaps immediately upon his arrival in 1476. The little property between the Chapter House buttresses was doubtless no more than what the Sacrist's accounts call it, a shop, which Caxton used only as a branch for booksales to church, nobility and law. He needed much larger premises to provide living quarters for his wife Maud, his daughter Elizabeth, and his employees, a workshop for his printing presses and equipment, and the main bookshop of the Red Pale, to which the *Advertisement* invited customers for the *Ordinal*.

Once again Tanner has identified the exact spot. The Almonry gate, with the loft above and the house adjoining it to the south, stood on the modern corner of Great Smith Street and Victoria Street, where the Department of Trade and Industry glasshouse now begins. The loft, at the low annual rent of 3s 4d, measured only eighteen feet by seven, and would surely be too cramped except for use as a store room; but the house, costing 13s 4d a quarter or

[1] This is why the Almoner went on entering the gate-loft until 1487, when the lease expired.

[2] Or rather, the missing Prior's accounts from 1457 to 1482 were probably entered in some book or books now lost, until in 1482 the new Prior Essex took over again the book left half empty by Prior Walsh in 1456, which now alone survives. Essex's account book also contains an undatable notice (except that it must have been made between 1482 and 1491) of five payments totalling £22 made to him on month-days including 23 February, 7 May, and 1 July in an unspecified year by 'Johannes Essalen [?] goldsmith', through Caxton's hands (*per manus W. Caxtoni*). Caxton's role seems to be that of an obliging friend rather than of a business participant in the transaction, whatever it may have been. I am grateful to Mr. Howard M. Nixon for fuller information on this note, which Crotch garbles, and for the confirmation that it is in Essex's own hand.

IV Caxton's Type 2, from Dicts of the Philosophers, 1477.

Here it is so that euery humayn Creature by the
ſuffrance of our lozd god is bozn z ozdeigned to
be ſubgette and thral vnto the ſtozmes of foztune
And ſo in diuerſe z many ſondzy Wyſes man is perplex-
id With Wozldly aduerſitees/Of the Whiche I Anto me
Wydeuille Erle Rpuyeres/lozd Scales zc haue largely z
in many diffirent maners haue had my parte And of hem
releued by thynfynyte grace z goodnes of our ſaid lozd
thurgh the meane of the Mediatrice of Mercy/Whiche hee
euidently to me knoWen z vnderſtonde hath compelled me
to ſette a parte alle ingratitude/And dzoof me by reſon z
conſcience as fer as my Wrecchednes Wold ſuffyſe to gyue
therfore ſynguler louynges z thankes/And exozted me to
diſpoſe my recouerd lyf to his ſeruyce/in foloWig his laWes
and commandemets/And in ſatiſfaccon z recompence of myn
Inyquytees z faWtes before don/to ſeke z execute y Werkes
that myght be moſt acceptable to hym/And as fer as myn
fraylnes Wold ſuffre me I reſted in that Wyll z purpoſe
Duzyng that ſeaſon I vnderſtode the Jubylee z pardon to
be at the holy Apoſtle Seynt James in Spayne Whiche
Was the yere of grace a thouſand.CCCC.lxxiij.Théne
I determyned me to take that Voyage z ſhipped from ſou-
thampton in the moneth of Juyll the ſaid yere/And ſo
ſayled from thens til I come in to the Spaynyſſh ſee there
lackyng ſyght of alle londes/the Wynde beyng good and
the Weder fayr/Thenne for a recreacon z a paſſyng of tyme
I had delyte z axed to rede ſome good hiſtozye And amóg
other ther Was that ſeaſon in my cópanye a Wozſhipful gen-
tylman callid loWys de Bretaylles/Whiche gretly delited

Iohannes De Gigliis alias de Iustis Apostolicus Et in Inclito Regno Anglie fructuum & prouentuum canonce & fixe debi? lop Collector / Et Decre? de Valuinis decanus Ecclie Sancti michael de leposto Honomen domini nostri pape Cubicularius Sedis apostolice Nuncii et commiffarii per eundem sanctissimum dominum nostrum papam adintra scripta deputati In pdicto anglie regno / Vniuerfis presentes litteras Inspecturis Salutem & sinceram in domino caritatem / Nouentis qp sanctissi? mus in cristo pater & dominus noster Innocentius diuina prouidentia papa octauus / & dominio hypteritie Ecclie ac arteis quibuscunqz uicti regni diction fubiectie qui per se vel aliu Intra tempz / ad issimi dni nri & sedis astine binplacitu duratura / & bisquicquo eiusdem bnplacitu renoctario aut offendoz in sius literis sufpensio facta fuerit sedm tenoze ipsaz literaz aplicaz / Qui ad impugnandum infideles & resistendu coz conatibz / Cantu Quatuor Cres vel Duos vel onu florenoz auri Vel bri qutum per nos Commiffarios prefatos defuper deputatos / seu cu Collectorib? a nobis fuper hoc ofcituendis vel ofcituendis etiam mendicantiu Regularis cuiatez Exitunquerent / & cu offectus perfoluerint / Ut Conf.lloz donec? prefbitez fecularis vel cuiufiu ordinis etiam Regularium fubueniunt fup hoc applica eez pciorum pfatos & crimibus eligendi / eligerie & eligentium roffeftione audita feu coffeftionib? efpertine audrtie pro commiffis quomodo cofulenda / Cofpiratoz In romanu Pontifice & in predictam fedem applicam / & inieetionis manuu pioletaz In Epos ec aliofue / & cu peccatis criminibz q occeffib? quibufcunqz qtticunqz enormibz / & grauibz /etiam fi talia foret propter que fedes applica eez hie femel in bitas & in alio dicte fedi no deferuatio ratib? & peccatis quocies id pecientibus auctoritate Aplica de abfolucionis biftica prouidere & tam femel in bita qp in mortis articulo plenariu oim fuoz ptoz remiffionum & abfolucioz cu ea plenaria Indulgentia qua de bere ac recuperatione terre fancte cofedem infidelium repugnation / ac Anno Iubileo que etiam ad pm obita & que alias aliis facerdotib cofeffi foret egtmdaf Iplis in fincritate fidei & bnitate fee Romane ecclie ac obediecia & deuocione sissimi dni nostri & fucceffoz fuoz Romanoz Pontificu Canonice intrancium perfiftentib? impendere Ita ut fi ipsis in hmoi mortis articulo & leptus cofticutos abfolucio ipfa impendat / Nichilomin? iterato in vero mortis articulo poffit impendi & impeza fuffragetur eifde aucto ritate aplica & aplice poteftatis plenitudine concebit facultatem prout in Ipsis litteris aplicis fuper hoc emanatis plenius continentur

Inter pfatam tpius dicti bnplacitu de facultatib? fuis Competentem quantitatem ad opus fidei hmoi ac ad epugnationem Infidelium Contulerit / Ideirco tenore prefentium hmoi Confessionis eligendi et Auctoritate apoftolica qua Jn hac parte fungimur fatiffacto tamen hiis quibus fuerit fatiffaccio impendenda plenariam ac liberam tribuim? facultatem/Datum Sub Sigillo Sancte Crucate Anno Incarnationis Diice Millesimo Quadringentesimo Octuagesimo Nono Diepneustime quarto Menfis Aprile

£2 13s 4d a year, was evidently substantial. The celebrated sign of the Red Pale unfortunately never turns up again in any document after its first and only appearance in the *Advertisement*; so we do not really know whether it was Caxton's personal sign or merely went with the premises (perhaps it had been Selley's trademark as a vintner?), nor whether Caxton used it only at this early period, or kept it all his life. The sign itself was no doubt a shield with a broad red stripe down the middle, or in heraldic language a 'pale'. Caxton kept the Almonry house and loft (as well as the Chapter House shop) until his death; from January 1485 he added the house next door at 13s 4d a year, payable quarterly (once, in July 1485, he paid in wine!); and from July 1488 he took a third adjacent house at £1 6s 8d yearly.

In 1480 Caxton made a number of modernising changes in his printing methods, straightening his line-endings, printing his quire signatures, and introducing a new small type 4 measuring only 95 mm. These features first appear together in *Chronicles of England*, 10 June 1480, and serve as a useful dividing line for putting undated books of this period in chronological order. Duff assumed that Caxton was spurred to these reforms by the opening in 1480 of the first London press by John Lettou. Lettou, a Lithuanian trained in Italy, printed in a neat small gothic type, 83G., which had been used at Rome in 1478-9 by Johannes Bulle from Bremen, with printed signatures and even line-endings. But Caxton was no doubt already well aware of the first Oxford press, which began on 17 December 1478 with a small Cologne type (97G., used at Cologne in the same year by Gerard ten Raem) and printed signatures; he was personally connected with the Schoolmaster Printer of St. Albans, who began in 1479 with a small bâtarde 89B. evidently made by Veldener; and Colard Mansion had adopted a moderate sized type (110GR.) in 1478 and even endings in 1479. In any case it seems certain that Lettou did not begin before the second half of 1480, several months after Caxton had completed his innovations.

Caxton's new type 4: 95B. was a reduced version of his type 2, with shorter side-flourishes and with a few majuscules (A, C, D, G, S) resembling his type 3. No doubt it was manufactured by Veldener as usual. With this much smaller type Caxton could now produce longer books without prohibitively larger paper costs, and resume his delightful career as a translator without being restricted to shorter texts. This, and not a mere desire to keep up with Lettou, was evidently his intention, when he ordered type 4 and planned his new production programme.

Caxton's decision to order type 4 must have been connected with the beginning of his vast *Ovid* translation and have occurred at the same time, towards the autumn of 1479. The French prose *Métamor- phoses moralisées*[1] retold Ovid's pagan love tales of gods and mortals

---

[1] The prose version was made towards 1460 from a French verse adaptation of the late fourteenth century from Pierre Bersuire's Latin *Ovidius moralisatus* (about 1330).

in the style of medieval romance, with moral explanations in terms of Christian allegory, so that the reader had it both ways. Caxton had produced no long prose texts since the *Recuyell*, but this was longer still, too costly for type 2, but practicable for type 4.[1] Whether Caxton ever printed his *Ovid* is an unsolved problem. Probably he found a patron, for the text survives in a two-volume presentation manuscript executed not by Caxton himself but by a professional scribe, with calligraphy and grisaille illuminations in Flemish Burgundian style.[2] But no trace of a printed edition has been found. Perhaps Caxton abandoned his project out of prudence; indeed, Colard Mansion was to be ruined outright by his own splendid French *Ovid* of May 1484, in his still larger type 162B. and with lavish woodcut illustrations, so that he had to flee Bruges and his creditors, and is heard of no more. On the other hand, in the prologue to *Golden Legend* about 1484 Caxton lists his translations in what is apparently their order not of translating but of printing, for *Ovid* appears after *Mirror of the World*, which he translated 2 January-8 March 1481 and printed soon after, and before *Godfrey of Boloyne*, which he translated 12 March-7 June 1481, and finished printing on 20 November 1481. The interval between the printing of these two books provides a possible gap, and the only suitable one, for an edition of the *Ovid*. Other large Caxtons (notably *Morte d'Arthur* and *Four Sons of Aymon*) exist only in one or two copies; and post-Reformation attitudes would give an adequate explanation for destruction of the *Ovid*, and the absence of sixteenth-century reprints. In any case, although type 4 had already arrived before Caxton finished translating *Ovid* on 22 April 1480, he could not print the book immediately, for *Chronicles of England* must have been already at press.

Two short works in type 4 are probably earlier than *Chronicles*, as they still lack printed signatures, and perhaps belong to the spring of 1480. The quarto *Officium visitationis B.V.M.*, intended for insertion in a manuscript breviary, gives the service for the Visitation of the Blessed Virgin on 2 July in the new form promulgated by Sixtus IV on 1 January 1475 to celebrate his rebuilding of Santa Maria del Popolo in Rome. The Archbishop of Canterbury, Cardinal

---

[1] The 220,000 words of Caxton's *Ovid* would have occupied about 400 leaves in type 2, about 230 in type 4. Paper costs made up about half the total cost of printing, and a substantial saving of paper would make all the difference between profit and loss.

[2] The second volume belonged to Samuel Pepys the diarist, who left it with his library to Magdalene College, Cambridge. The first remained unknown until discovered by the brothers Lionel and Philip Robinson in the residue of the Sir Thomas Phillipps collection; it was sold at Sothebys on 27 June 1966 for £90,000, and was restored to its companion at Magdalene partly by the heroic efforts of the College and English sympathisers, but mostly through the generosity of three Americans: the bookdealer L. D. Feldman who renounced his rights as buyer, the head of University Microfilms Eugene B. Power who lent the money, and George Braziller who published a colour facsimile edition in 1968 and devoted profits to the rescue fund.

Thomas Bourchier, received proposals in 1480 to make the Visitation feast obligatory. Caxton probably produced his edition well in advance of 2 July and under commission, by way of propaganda for the new office. But the Sixtine feast apparently did not gain currency in England, for later editions printed by William de Machlinia about 1485 and Pynson about 1496 are in the older form instituted by Urban VI in 1389, and the Urban form alone is found in Sarum, York, and Hereford breviaries.[1]

Doctrine to learn French and English is a vocabulary and phrase-book for schoolboys, merchants, and travellers, adapted into this English form about 1465-6 (as P. Grierson has shown from monetary and currency references) from a French-Flemish *Livre des métiers* compiled at Bruges soon after 1367.[2] This is one of the most enter-taining of all Caxton's books, and a fundamental source for colloquial English in the late fifteenth century. Caxton was not the translator, as many have assumed, for the diction and even the spelling habits are quite unlike his. However, he did add a few words in French and English in the colophon: 'at Westminster by London in forms em-printed', and perhaps inserted a little other English matter, such as the list of fairs held in England ('To the fair of Stourbridge [at Cambridge], to the fair of Salisbury, to Saint Bartholomew's fair which shall be at London'). Perhaps Caxton knew the translator, who was no doubt an English merchant in Flanders at the time of the Utrecht exodus, and perhaps he brought the *Doctrine* with him to England for the sake of its racy humour, which is right up his street. The book's unique feature is a series of bilingual character sketches, evidently not wholly fictitious, of men and women, servants, shop-keepers and craftsmen, by a satirical narrator. 'Colombe the halting [i.e. lame] went her chiding from hence for this, that I would have kissed her, nevertheless I had no lust, and she me cursed, and I cursed her again.' 'George the book seller hath more books than all they of the town. He buyeth them all such as they be, be they stolen or emprinted,[3] or otherwise purchased [i.e. irregularly come by]. He

---

[1] Caxton's *Officium visitationis* survives only as a fragment in the British Library comprising the last seven leaves. A. I. Doyle has shown that this once belonged (until removed at some time between 1763 and 1807) to the West-minster Abbey scribe William Ebesham's tract-volume now in the John Rylands Library which also includes Caxton's *Propositio Johannis Russell*. Comparison with the Roman *Breviary* of the time (e.g. the edition of P. de Plasiis and B. de Blavis, Venice, 18 April 1479) shows that a complete copy would occupy three quires, totalling twenty-four leaves.

[2] *Doctrine* is Caxton's title. '*Vocabulary*' (Blades and Duff) or '*Dialogues*' (H. Bradley) are modern inventions. Duff mistook *Book to learn to speak French* printed by Pynson and Wynkyn de Worde in the middle 1490s for a reprint of Caxton's edition, whereas it is an entirely different text intended for trade with France rather than Flanders.

[3] This would be the first allusion to printing in the English language if written in 1465-6; but perhaps Caxton altered it in 1480 to amuse himself. The original French must have been '*empruntés*', i.e. borrowed, the point being that George's stock was ill gotten.

hath Doctrinals, Catos, Hours of Our Lady, Donatuses, Parts of Speech, Accidences, Psalters well illumined, bound with clasps of silver, books of physic, Seven Psalms, calendars, ink and parchment, pens of swans, pens of geese, good Portoses [i.e. Breviaries] which be worth good money.'[1] 'Bring us to sleep, we be weary,' say the travellers, and 'Well, I go, ye shall rest,' declares the hostess, 'Jenette, light the candle, and lead them there above, in the solaire tofore, and bear them hot water for to wash their feet, and cover them with cushions.'

Meanwhile Caxton continued to use type 2* concurrently with type 4. Laurentius de Saona was so pleased in 1479 with his printed *Nova Rhetorica* that he lectured on it in Paris that winter, completed a shortened version there called *Epitome Margaritae eloquentiae* on 24 January 1480,[2] and commissioned Caxton to print this also. Caxton's edition, a folio of thirty-four leaves, is still without printed signatures, but has even line-endings; so, if (as is likely) it came just before *Officium visitationis*,[3] it was perhaps his first book with this feature.[4] But a second edition of *Dicts*, again without printed signatures, is perhaps his earliest book of 1480, for it shows a transitional stage, with most endings even but not all. This is a page for page reprint, made from a copy with the variant colophon, without bothering to change the date 18 November 1477. Evidently Caxton followed the practice, which examples from other printers show to have been normal, of keeping for future use a copy of the last and therefore correctest impression to leave the press. Someone thought it worth while to make a manuscript copy of *Dicts II* (now British Library MS. Add. 22718), with a separate blank quire on which births in the family of Hill, of Spaxton, Somerset, were entered, the earliest being of 30 May 1479; but this entry was no doubt made retrospectively, and cannot be used in evidence, for the line-endings hardly allow a date for *Dicts II* earlier than the beginning of 1480.

Abbot Sant of Abingdon, in one of the bouts of financial in-fighting which were so inevitable among rival Indulgence com-missaries, had landed himself in trouble with Pope and King. Sixtus IV revoked his licence on 15 December 1478 and instructed Johannes de Gigliis to look into his accounts. Edward came down on him for issuing the 1476 *Indulgence* without royal permission, but gave him a pardon on 16 June 1481.[5] Caxton, adaptable as ever, was

[1] Although the list of George's stock was drawn up before the invention of printing, Caxton himself printed almost all these books.

[2] Misprinted 21 January by Caxton. The author's manuscript (Vatican Latin 11441) says 24 January.

[3] *Officium* has even endings, *Doctrine* has a text arranged in phrases of varying length, so one can't characterise its endings. Both these books contain a remnant of paper with a mermaid watermark, made in Champagne.

[4] The only known copy of *Epitome* was first identified as a Caxton by Mrs. Jean E. Mortimer in Ripon Cathedral Library in 1953, and is now in the Brotherton Library at Leeds University.

[5] However, Sant and Gigliis were sent to Rome together by Edward in 1479 to support his claim to interfere in Italian affairs, so Sant was not in disgrace *qua* diplomat nor altogether at odds with Gigliis.

equally at home with the new commissary John Kendale, 'Turci-pelerius' or honorary infantry commander of the Knights of St. John of Jerusalem at Rhodes, who collected for defence of the island against the besieging Turks; indeed, Caxton and Kendale witnessed the churchwardens' accounts of St. Margaret's Westminster to-gether in May 1480, just before Kendale set off to collect in Ireland. The only known copy of Caxton's first edition of Kendale's *Indul-gence*, in nineteen even-ended lines of type 2*, was sold to Simon 'Mountfort' and his wife Emma on 31 March 1480.[1] It begins with a four-line initial F, apparently of Dutch or Flemish origin, which does not occur elsewhere in Caxton's work. This is his first and only use of a woodcut initial until he acquired his first complete set for use in the 1484 *Aesop*. Soon Lettou produced two more editions of Kendale's *Indulgence* in his neat small type, and Caxton, not to be outdone, printed another in type 4, which must have appeared after 7 August 1480, for it includes (as does also a third edition from Lettou) an additional line of text giving the date as Sixtus's tenth papal year, which began on that day. Next year Gigliis himself was commissary for a new *Indulgence*, occasioned by the Turks' latest and furthest penetration into the West, the capture of Otranto in the heel of Italy in 1480. Caxton printed two editions of this (one a 'single issue' for individual purchasers, the other a 'plural issue' for two or more purchasers) dated in Sixtus's eleventh year, i.e. after 7 August 1481.

Meanwhile *Chronicles of England* had been completed on 10 June 1480. The French original, *Brut d'Angleterre*, gave a history of England from the mythical times of Brutus the Trojan to 1333, and was translated into English later in the fourteenth century with a continuation to 1377. Two further continuations, both included in Caxton's edition, covered the periods to 1419 and 1461. Blades airily suggested that Caxton himself wrote the 1419-61 supplement; but this was in fact compiled soon after 1464 (as references to Mar-garet of Anjou's exile in Lorraine and to the death of Pope Pius II on 14 August 1464 reveal), well before Caxton began authorship, and the best surviving manuscripts show that he printed from an already corrupted text. The continuation shows proper Yorkist sentiments, but had to stop at 1461, for—although chronicles con-tinued to be written for private circulation and posterity during Edward's reign—everything after Edward's coronation was still dynamite. This is the work, as robustly and humorously English as Chaucer, which we have quoted on incidents from Caxton's youth. Every Englishman at that time learned his national history from it, and a century later, by way of its Lancastrianised embodiments in later chronicles such as Hall's and Holinshed's, it became Shakespeare's

---

[1] Another copy, sold on 16 April 1480 to Richard and John Cattlyn, was reported to William Herbert in the 1780s, but has not been heard of since. Mountford was beheaded on 3 February 1495, along with William Daubeney, another patron and friend of Caxton, for complicity with Perkin Warbeck.

source for the tragedies of Lear and Cymbeline, and the English historical plays up to Henry VI.[1]

Caxton printed *Chronicles of England* 'at request of diverse gentlemen', and added a brief epilogue praying for 'a very [i.e. true] final peace in all Christian realms, that the infidels and mis-believers may be withstanden and destroyed and our faith enhanced, which in these days is sore diminished by the puissance of the Turks and heathen men'. This remark was no doubt intended to please the papal friends who had commissioned his *Indulgences*, for Sixtus was then pressing Edward to give money and men for a crusade against the Turks. Caxton also prayed God, no less topically, to send Edward 'the accomplishment of the remnant of his rightful inheritance beyond the sea'. Thoughts of war with France had in fact revived in that summer, as part of a possible deal with Maximilian in which Duchess Margaret was deeply interested. Margaret visited England from late June to September 1480 in the hope of influencing Edward, who put her up in the sumptuous though uncomfortably named residence for visiting V.I.P.s called Coldharbour in Thames Street. Probably Caxton's helpful words were written in knowledge of her intention and early arrival, and he saw his 'right redoubted lady' for the last time, and to his profit, during her stay. The suggestion of a Sotheby cataloguer, that the *Ovid* manuscript was executed for presentation to Duchess Margaret at this time, seems very likely, and the possibility that the *Ovid* was actually printed not long after, towards mid-1481, is reinforced if she was indeed its patroness. On her way home Rivers entertained Margaret and the King at his Kentish manor, the Mote near Maidstone, which Warwick's gangsters had tried to sack in 1468, the year of her marriage.

On 18 August 1480 Caxton completed an edition of *Description of Britain*, extracted from the first book of Trevisa's translation of Higden's *Polycronicon*,[2] and intended as a supplement to *Chronicles*, because, as he remarks in his prologue, 'it is so that in many and diverse places the common Chronicles of England be had and are now late emprinted at Westminster'. Nearly all surviving copies are in fact bound either with the first edition of *Chronicles* or with the

---

[1] Kingsford's statement, based on Brie's Early English Text Society edition, that Caxton's text contains 'a gap covering nearly the whole of 1458 and 1459', has never been refuted. In fact Brie was here using an imperfect British Museum copy (shelfmark IB. 55062) of Caxton's second edition, 8 October 1482, without realising that it lacked two leaves (y2, y5), the text of which is of course present in perfect copies of both editions. Caxton no doubt particularly enjoyed printing an entry which the *Chronicles* had borrowed from Rolewinck's *Fasciculus temporum*: 'Also about this time (1457) the craft of emprinting was first found in Magounce (Mainz) in Almayne, which craft is multiplied through the world in many places and books be had great cheap and in great number by cause of the same craft.'

[2] Higden-Trevisa, however, describes Ireland, Scotland, Wales, England in that order. Caxton, with his typical editorial zeal spurred on by patriotic feeling, rearranges the sections in the reverse order, England, Wales, Scotland, Ireland.

second edition which Caxton produced on 8 October 1482. Curiously enough, the book has no printed signatures, perhaps because he felt it really formed part of another work; in any case, this is positively the last Caxton which lacks them.

No dated book occurs in the next half year, and no undated book seems typographically suitable for assignment within this gap. It may be that Caxton began work on the lost *Ovid* (if it ever existed) at this time, and postponed its completion—as he apparently did with another large project, the *Canterbury Tales* of 1476—to perform more urgent commissions, or to await the completion of financing by its patron.

12

FRIENDS IN COURT

CAXTON translated *Mirror of the World*, an illustrated hand-book of popular science, from 2 January to 8 March 1481, and no doubt printed it soon after. He did so 'at the request, desire, cost and dispense of the honourable and worshipful man Hugh Bryce Alder-man and citizen of London, intending to present the same unto the virtuous, noble and puissant lord William Lord Hastings, Lord Chamberlain unto the most christian king Edward the Fourth, and Lieutenant for the same of the town of Calais and Marches there'.[1] Hugh Bryce, the Irish goldsmith whom Blades miscalled a Kentish mercer, was a former colleague (from 1469 onwards) and Hastings a former chief (from 1465) in the trade negotiations at Bruges; and Bryce himself as a Governor of the Royal Mint in the Tower was a subordinate of Hastings as Master of the Mint. Caxton, always rather pushing, felt entitled to address Hastings directly ('I humbly beseech my Lord Chamberlain to pardon me of this rude and simple translation'), and even brought in the King, 'under the shadow of whose noble protection I have emprised and finished this little work and book, beseeching Almighty God to be his protector and defender again all his enemies and give him grace to subdue them and in especial all them that have late enterprised again right and reason to make war within his realm.' In May 1481 war had broken out against Scotland, in response to the land and sea raids which Louis had instigated during the past twelve months. Hastings was fitting out a fleet at Calais with picturesquely named big guns (including the Great Edward, the Little Edward, and Fowler of Chester) and 180 brass hand guns. Edward was raising an army to cross the Border (though in the end he left all the fighting to Gloucester). Caxton's topical remarks show that *Mirror of the World* was printed at this time.

Caxton translated from a later French prose version of the French verse *Image du monde* written in 1245-6 by a certain Gossouin of Metz, using a manuscript which has been generally but wrongly identified with British Library MS. Royal 19.A.ix. The latter belongs to the same group as Caxton's, originating at Bruges with a manu-script written in June 1464 to the order of Jehan Le Clerc bookseller, but omits various illustrations and the manuscript captions to the

---

[1] Caxton's words do not preclude his having chosen to translate *Mirror* of his own accord before obtaining Bryce's commission. In view of the book's connec-tions with Bruges and the 1460s this is not unlikely; though of course it is possible that Bryce himself had acquired the manuscript at Bruges.

in lyke wyse on
bothe sides, and
that eche of them
threwe a stone in
to the hole. Whe-
ther it were gre-
te or lytyl / eche
stone shold come
in to myddle of
therthe, wythout
euer to be remed
uid fro thens /
But yf it were
drawen away by

force. And they sholde holden them one aboute another
for to take place eueriche in the myddle of therthe. And
yf the stones were of like weight, they shold come therto
alle at one tyme, assone that one as that other. ffor nature wold suffre it none other wise. And that one shold
come ayenst another as ye may playnly see by this fy-
gure /

a   Nor yf their weyght and powers were not egall fro
        the place fro whens they shold falle / that whiche
were most heuy, that sholde sonnest come to the myddle of
therthe. And the other shold be al aboute her, as this se-
conde figure sheweth playnly on that other side /

x   Nor so moche may be caste therin that the holes
        may be full, lyke as they were to fore. As ye may

2 Caxton's Type 2* from Mirror of the World, 1481.

diagrams which are all present and correct in Caxton's edition;[1] so Caxton must have used a different and better manuscript, now lost. Caxton's amusingly and unashamedly incompetent woodcuts are the first printed illustrations in any English book; perhaps this was the special feature which attracted the patronage of Bryce and Hastings.[2] One of the most delightful displays medieval man's awareness of the spherical world (that he thought it flat is merely a modern popular fallacy). At the round earth's imagin'd corners stand four jaunty little men, two sidelong, one upside down, each about to drop a huge stone down a hole reaching to the centre, where six previous stones are seen duly suspended.

Not all the information in *Mirror* was so reliable. Caxton conscientiously omitted an assertion that Englishmen have tails, inserted mentions of Oxford and Cambridge as seats of learning and Bath as a health resort, and added eyewitness rectifications on the celebrated St. Patrick's Purgatory, a cave on an island in Lough Dergh on the River Shannon in western Ireland, where pilgrims were believed to see horrific visions of the next world. 'I have spoken with diverse men that have been therein, and that one of them was a High Canon of Waterford which told me he had been therein five or six times, and he saw nor suffered no such things. He saith that with procession the religious men that be there bring him into the hole and shut the door after him, and then he walketh groping into it where as he said be places and manner of couches to rest on. And there he was all the night in contemplation and prayer, and also slept there, and on the morn he came out again. Otherwhile in their sleep some men have marvellous dreams, and other thing saw he not. And in like wise told me a worshipful knight of Bruges named Sir John de Banste that he had been therein likewise and see none other thing but as afore is said.'

Mirror began a new series of texts of medium length in type 2*, and was followed by *Reynard the Fox*, Caxton's only translation from Dutch, which he finished on 6 June 1481 and no doubt printed immediately after. Only the next day, 7 June 1481, he also finished translating a much longer work, *Godfrey of Boloyne*, which he had begun on 12 March 1481, only four days after finishing *Mirror of the World*. Evidently he must have translated the greater part of the uncommissioned *Reynard* before he was obliged to begin the commissioned *Mirror* on 2 January 1481, and probably in the otherwise unoccupied autumn of 1480; and on 6 June 1481 he can have done little more than add the brief dated epilogue to *Reynard* and pass the work for press. It has been suggested that the word 'finished' in Caxton's colophon to *Reynard* refers to the completion of printing, not of translation.

---

[1] The manuscript captions are written in the same hand in all known copies of Caxton's edition; but this is doubtless the hand of one of his employees and not, as sometimes supposed, Caxton's autograph.

[2] However, it is likely that Bryce's gift to Hastings was not a copy of the printed edition, but a presentation manuscript of Caxton's translation.

Caxton does indeed use 'finished' in either sense according to context, and *Reynard* is not the only book for which scholars have been tempted to misinterpret as a printing date what is really only a translation date. In fact Caxton's practice is absolutely consistent and unambiguous. He uses the formula 'emprinted and finished' or the like to mean completion of printing, and 'translated and finished' to mean completion of translation. In two books only—*Game of Chess I*, and *Polycronicon*—the word 'finished' is used by itself; but here likewise the context, which concerns solely matters of writing and does not mention printing, shows that the date relates to completion of text, not to printing. The *Reynard* colophon says: 'by me William Caxton translated into this rude and simple English in the Abbey of Westminster, finished the vi day of June the year of Our Lord MCCCClxxxi and the xxi year of the reign of King Edward the iiiith'. This evidently belongs to the 'translated and finished' category, and the date must refer only to completion of translation.

Caxton's 'copy which was in Dutch' is now generally identified as the printed edition produced by Gerard Leeu at Gouda on 17 August 1479. This may be; but no conclusive or even positive evidence has been produced, and it seems equally possible that Caxton used a now lost manuscript closely related to Leeu's. Caxton's *Reynard the Fox* has never ceased to be popular, and remains a familiar children's book to this day; it is because of *Reynard* that we still call a bear Bruin and a cat Tib, and that the French for fox was changed from *goupil* to *renard*. The Dutch prefacer apologises for the immorality of this satire on the cunning and folly of court life in the palace of King Noble the Lion, where the animals behave so like their human counterparts; he means only to expose the 'subtle deceits that daily be used in the world', and 'who that may read therein though it be of japes and bourds [i.e. jests] yet he may find therein many a good wisdom and learnings'. Caxton adds a joke of his own: 'if anything be said or written herein that may grieve or displease any man, blame not me but the fox, for they be his words and not mine'. A curious typographical mishap occurred at the end of quire h, where the compositor omitted a page of text and was obliged to insert an extra leaf containing the matter of half a page on each side. One can see just how the accident happened. The previous page and the inserted leaf both end with the word 'for', and this 'homoeoteleuton' evidently caused the unwary compositor to leave out the equivalent of a page of text previously marked by the caster-off.

Immediately after printing *Reynard* in June-July 1481 Caxton began the production of his next book, a collection of three classical dialogues in English translation, Cicero's *De senectute* and *De amicitia* and Bonaccursius de Montemagno's *De vera nobilitate*.[1] He gives the date of printing only for the first of these, 12 August 1481, and the

---

[1] Colard Mansion had printed *Bonaccursius* at Bruges before June 1477 in the French version by Jean Miélot.

other two, which share a separate set of signatures, may either have followed a few weeks later, or have been printed concurrently on another press. *De senectute*, or '*Tully of Old Age*', was translated as Caxton correctly says at the order of 'the noble ancient knight Sir John Fastolf of the county of Norfolk banneret[1] living the age of four score year', from the French version (completed on 5 November 1405) by Laurent de Premierfait. Sir John Fastolf (1378-1459), although he was to become a secondary original of Shakespeare's fat knight,[2] had fought in the French wars with unblemished prowess and temperance; he was a patron of the Pastons, and stepfather of Stephen Scrope (c. 1399-1472), the same whose translation of *Dicts* had remained unknown to Rivers and Caxton. As C. F. Bühler has suggested, Scrope may well have been the translator of *Tully of Old Age* also; Scrope's text, as revised by William Worcester the former secretary of Fastolf, may have been the 'Tully de senectute translated by me into English' which Worcester presented to William Waynflete Bishop of Winchester at Esher on 10 August 1473 ('but I received no remuneration from the Bishop', Worcester ruefully adds); and it was doubtless the one listed, along with Caxton's *Game of Chess I*, among Sir John Paston's English books in 1475-9.[3] Possibly Caxton obtained his manuscript through Ebesham; but if so, this must have been a private transaction for which Ebesham made him pay through the nose, and not a positive commission from the Paston family, for Caxton complains feelingly that it was 'with great instance, labour and cost comen into mine hand'.

As for *Of Friendship* and *Declamation of Nobility*, Caxton attributes their translation to John Tiptoft, Earl of Worcester, whom Warwick had beheaded on 18 October 1470; but here a baffling mystery arises. Sir John Paston's book list also includes a 'Tully de Amicitia left with William Worcester'. May not this also, as Bühler suggests, have been translated by Scrope and revised under his own name by William Worcester? May not Caxton, or whoever supplied his manuscript, have misidentified this Worcester in good faith as the more famous Earl? On the other hand, Caxton, with his intimate contacts among the Yorkist magnates, is perhaps more likely to have been correctly informed. Another piece of evidence seems to point more decidedly to Tiptoft. It has been generally assumed that *Of Friendship* was translated, like *Of Old Age*, from Premierfait's

---

[1] A banneret was a knight entitled to enlist vassals under his own banner, or one knighted for his prowess in the King's presence on the field of battle.

[2] There are real points of resemblance between Falstaff and Fastolf, who was brought up in the household of the Duke of Norfolk, owned a tavern in Southwark called the Boar's Head, and was accused though unjustly of cowardice at the Battle of Patay in 1429, a distressing incident in which he appears under his own name in *Henry VI part 2*, act 3, scene 2; act 4, scene 1.

[3] Sir John Paston's copy of *Game of Chess I* is in all probability one now in the British Library (shelfmark C.10.b.23). Paston may well have met Caxton in Bruges at Duchess Margaret's wedding in 1468, and later during his service under Hastings at Calais from 1473 onwards.

French version, which is often included in manuscripts of his version of *De senectute*, and that *Of Nobility* was translated from the French version made by Jean Miélot for Philip the Good in 1449. If the two Cicero works really derived from the same French manuscript, it would indeed be difficult to argue that they had two different English translators each named Worcester! But I have found by comparing the English *Of Friendship* and *Of Nobility* printed by Caxton with the French of Premierfait and Miélot that the divergences are so complete as to make it impossible that the French versions could have served as basis for the English. Indeed, both pieces in English are so close to the Latin originals that they can only have been translated directly from the Latin. William Worcester was a notoriously poor Latinist, but Tiptoft was among the most distinguished of English fifteenth-century humanists, having learned Latin at the University of Padua and drawn tears from Pope Pius II by the beauty of his Latin eloquence.[1] Caxton himself implies that he acquired his second and third pieces separately from the first, and does not specify, as he always does when this is the case, that they were translated from French. Perhaps it still remains more likely that Tiptoft was the translator; in any case, Caxton believed he was.

However this may be, Caxton took the opportunity to make a panegyric of the execrated long-dead Tiptoft, 'to whom I knew none like among the lords of temporality in science [i.e. learning] and moral virtue', praising his 'great labours in going on pilgrimage to Jerusalem', exclaiming 'what worship had he at Rome in the presence of our Holy Father the Pope,[2] and so in all other places unto his death, at which death every man that was there might learn to die and take his death patiently, wherein I hope and doubt not but that God received his soul into his everlasting bliss'. In fact Tiptoft, hoping to make a better impression in the next world, had asked his executioner to behead him with three blows, 'in honour of the Trinity'. The crowd had tried to lynch him in the Strand, the most unpopular man in the realm, Edward's obedient and merciless judicial killer since 1462, called the Butcher of England. He had not only hanged, drawn and quartered the Warwick men captured by Rivers at Southampton in April 1470, which was fair enough, but impaled them as well in his un-English way, using as men said 'the law of Padua, not the law of England'. Who tipped off Caxton to rehabilitate Tiptoft, to commend him for the same merits of pilgrimage, polite learning, 'virtuous disposition' and charming the Pope for which he had praised Rivers in *Cordiale*? Caxton's readers would understand the hidden message, that a living man shared the martyred Butcher's virtues and deserved his advancement, and guess

[1] Tiptoft, however, did not translate *Caesar's Commentaries as much as concerns this realm of England*, 1530 (STC 4337) as often said.

[2] Tiptoft's grand tour lasted from 1458 to 1461. He sailed to Jerusalem on the spring pilgrimage galley from Venice on 17 May 1458, returned on 6 September, charmed Pius II in 1460, and landed in England again on 1 September 1461.

that Anthony Earl Rivers was he. No one had better cause to regret Tiptoft, the Woodville protégé who perished in their defence against Warwick, or to covet his great post as Constable, once shared by Rivers father and son but now held by Gloucester,[1] or to need power against a new Warwick to shield his ward and nephew the young Prince. Gloucester in that August was in the North, the coming hero, preparing to fight the Scots; Rivers's marriage with Margaret of Scotland was in abeyance till after the war, and Edward's army, in which Rivers was appointed to command a thousand men, never set out. So Caxton's praise of Tiptoft was undercover propaganda for the dissatisfied Rivers, who doubtless rewarded him. But Caxton also declared that *Of Old Age* was particularly suitable for 'noble, wise and great lords, gentlemen and merchants', especially 'them that be past their green youth', while the whole volume was printed 'under the umber and shadow of the noble protection of . . . King Edward the Fourth, to whom I most humbly beseech to receive the said book of me William Caxton his most humble subject and little servant, and not to disdain to take it of me so poor, ignorant and simple a person, and of his most bounteous grace to pardon me so presuming'.

It is within the rather brief interval between the *Tully* of 12 August 1481 and *Godfrey of Boloyne*, 20 November 1481, that we must fit Caxton's lost edition of *Ovid*, if we are willing to agree that it existed, and to accept the evidence of the *Golden Legend* prologue that it was completed after *Mirror* and before *Godfrey*. If so, only the closing section of *Ovid* can have been actually printed within those three months, in which the 140,000 words of *Godfrey* would not have left room for all the 220,000 words of *Ovid*; but we can suppose if we wish that the greater part of *Ovid* was printed during the otherwise empty end of 1480, and the rest postponed owing to the *Mirror* and *Tully* commissions. *Ovid*, if printed, would certainly be in type 4 like *Godfrey*.

Caxton had translated *Godfrey of Boloyne* from 12 March to 7 June 1481, from a French version of Archbishop William of Tyre's Latin history of the First Crusade, using only books i-ix, which end with Godfrey's death in 1100. His prologue is more than usually expansive concerning the book's place in his publishing programme, its topical interest and propaganda purpose, and its patrons. Godfrey, he explains, is the last but not the least of the Nine Worthies, whom

---

[1] Tiptoft was Constable of England from February 1462, when he beheaded Oxford's father, to August 1467, when (soon after organising Anthony's Smithfield tournament against the Bastard of Burgundy) he was sent to Ireland at the Queen's wish to liquidate her enemy Desmond, and handed over his post as Constable to Rivers the father, evidently as part of the deal. Father Rivers held it till beheaded by Warwick in 1469, and Anthony was supposed to have inherited it in theory until Tiptoft regained it in March 1470. To make matters more complicated, Scrope the translator of *Old Age* was a cousin of Tiptoft the translator of *Friendship*; Scrope's mother Millicent, who married Sir John Fastolf in 1409, was the widow of Sir Stephen Scrope and a daughter of Robert, 3rd Lord Tiptoft, the elder brother of Tiptoft's father.

he lists: the ancients Hector, Alexander, and Julius Caesar; the Hebrews Joshua, David, and Judas Maccabaeus; and the Christians King Arthur, Charlemagne, and Godfrey. Caxton's enthusiastic accounts of Arthur, 'so glorious and shining'—'of whose acts and histories there be large volumes and books great plenty and many, o blessed Lord, when I remember the great and many volumes of Saint Graal, Galahad, and Lancelot de Lake, Gawain, Perceval, Lyonel, and Tristram!'—and of Charlemagne, surely mean that he had already resolved to print their stories, as he did four years later in 1485. As for Godfrey, 'whose noble history I late found in a book of French', the events which occasioned his deeds are, 'as me seemeth, much semblable and like unto such as we have now daily tofore us by the misbelievers and Turks emprised against Christendom'; for the Turk 'now late this said year hath assailed the city and the castle in the Isle of Rhodes, where valiantly he hath been resisted, but yet notwithstanding he hath approached more near and hath taken the city of Ydronte in Puylle [Otranto in Apulia], by which he hath gotten an entry to enter into the realm of Naples, and from there without he be resisted unto Rome and Italy[1] . . . then me seemeth it necessary and expedient for all Christian princes to make peace, amity and alliance each with other and provide by their wisdom the resistance against him for the defence of our faith and mother Holy Church'. Who could be fitter to make first move than Edward, 'under the shadow of whose noble protection I have achieved this simple translation, that he of his most noble grace would address, stir, or command some noble captain of his subjects to emprise this war against the said Turk . . . And for to deserve the tenth place [among the Worthies] I beseech Almighty God to grant and octroye [vouchsafe] to our said sovereign lord or to one of his noble progeny, I mean my Lord Prince and my Lord Richard Duke of York, to whom I humbly beseech at their leisure and pleasure to see and hear read this simple book, by which they may be encouraged to deserve laud and honour.'

Caxton's propaganda was calculated to please many great persons, and perhaps each of these rewarded him. Pope Sixtus IV had circularised all Christian rulers to forget their quarrels and join the crusade. Johannes de Gigliis was in England selling his 1481 *Indulgence* for aid to Rhodes and Otranto, of which Caxton had just printed two editions. Edward had no intention of crusading, but made encouraging noises to placate Sixtus, blaming the perfidy of the Scots.[2] Rivers would be gratified by the build-up for his little

---

[1] Caxton writes before 25 March 1481, when by his calendar the year 1480 ended. Otranto was captured by the Turks in August 1480, and the news of its recapture by King Alfonso of Naples in August 1481 no doubt reached him after his prologue was already in print.

[2] It was not until September 1481 that Edward at last allowed Sir John Weston, Prior of the Knights of Rhodes in England, and John Kendale, for whom Caxton had printed the 1480 *Indulgence*, to leave for Rhodes with his royal letters of protection to the Grand Prior, which he had issued in April 1480.

Prince and the reference to the 'noble captain', who in the context could be no other but himself. Caxton's outright, twice repeated dedication of his book to the King is unique, although he had already produced three books (*Jason, Mirror, Tully*) 'under the shadow of his noble protection'. Perhaps the very copy he presented to Edward survives; it belonged later to the mercer Roger Thorney, a patron of Wynkyn de Worde and probably a friend of Caxton, and contains the inscription: 'This was Edward the Fourth book.'[1]

In August 1482 Gloucester captured Berwick (which Margaret of Anjou had handed over to the Scots in 1461 in return for aid against Edward) and entered Edinburgh, and the war was over. The King of Scotland again promised to send his sister Margaret to marry Rivers; but Edward decided against it, and Rivers married a not very brilliant and only distantly royal-blooded heiress instead, Mary Fitzlewis, a grandchild of Margaret of Anjou's favourite Edmund Beaufort Duke of Somerset. Mary of Burgundy fell from her horse and died on 27 March, and her husband Maximilian, denied all help by Edward, signed the Treaty of Arras with France on 23 December 1482. Edward's foreign policy of twenty years was in ruins, and he was left without an ally. The promised marriage between his eldest daughter Elizabeth and the Dauphin Charles was off. Hastings was teaming up with Gloucester's party, Buckingham, Arundel and the Stanleys, against the Woodvilles; he told the mayor of Canterbury that summer that the King was ill, and warned him to expect a revolution.

Caxton's long honeymoon with his patrons was nearly over. He produced only two books which certainly belong to 1482, a second edition of *Chronicles of England*, 8 October 1482, reprinted page for page from the first (a labour-saving method which made casting off unnecessary), and the huge *Polycronicon* (450 leaves, 430,000 words). *Polycronicon* was perhaps printed concurrently with *Chronicles II*, for Caxton dated his editorial epilogue 2 July 1482, and the finished volume was on sale no later than 20 November 1482, when a copy[2] was bought from Caxton in person by a certain William Purde, who calls him the King's Printer (*Regius Impressor*). In fact *Polycronicon* is the fifth and last book which Caxton states was produced 'under the noble protection' of Edward. His title as King's Printer, though doubtless genuine, has not been traced elsewhere, probably because it was still new, and because Edward had only a few months to live. Probably Caxton lost this post under Richard III; in any case, another man beat him to it soon after Bosworth Field, no doubt for political reasons.[3] Neither book has any

---

[1] Now in the possession of the Rosenbach Foundation, Philadelphia.

[2] Recorded in M. C. Tutet, Sale Catalogue (15 February 1786), no. 479, later acquired by Sir John Shuckburgh, at present still in private ownership.

[3] Peter Actors, native of Savoy, a London and Oxford bookseller but not a printer, was appointed King's Stationer for life on 5 December 1485, with licence to import 'books printed and not printed' free of duty. The printer

apparent patron, though possibly *Chronicles II* in 1482 was sponsored
by the same persons as *Chronicles I* in 1480—unless Caxton's compositor was merely mechanically following his copy when he reprinted word for word his master's remarks on 'request of diverse
gentlemen', Edward's 'rightful inheritance', and the Turkish peril.

Polycronicon, a world history from the Creation to 1360, was
written in Latin by Ranulph Higden, a Benedictine monk of Chester
(*d.* 1364), and translated into English in 1387 by John Trevisa
(1326-1412), vicar of Berkeley in Gloucestershire, at the command
of his patron Thomas, Earl of Berkeley.[1] Trevisa wrote robust
English with humorous comments after Caxton's own heart. The
philosopher Zeno 'said that man's soul shall die with the body': 'I
would a wise man had seen his water and poured it in his throat
though it had been a gallon,' exclaims the scandalised Trevisa. But
his diction was a century old, and Caxton 'somewhat changed the
rude and old English, that is to wit certain words which be neither
used nor understanden', so that the book was 'a little embellished
from the old making'. Caxton also added a continuation covering the
period 1358-1460, consisting almost entirely of the corresponding
section of his own edition of *Chronicles of England*, but with a few
additions to bring it up to world chronicle standard: 'I have not nor
can get no books of authority treating of such chronicles, except a
little book named *Fasciculus temporum*[2] and another called *Aureus de
universo*,[3] in which books I find right little matter since the said
time . . . If I could have founden more stories I would have set in it
more.' So Caxton in Edward's last year (and perhaps not by coincidence, for many besides Hastings knew the King was sick and an era
was ending) twice printed this Yorkist version of the calamities of
Lancaster, in which Edward's coronation was shown as the beginning
of a brave new world, 'humbly beseeching his most noble grace,' as
he wrote in the prologue to *Polycronicon*, 'to pardon me if anything be
said therein of ignorance or otherwise than it ought to be'. Whether
he foresaw it or not, this was Caxton's last address to Edward.

Just as in *Godfrey of Boloyne* Caxton had advertised his future

William Faques became King's Printer towards 1504, and was succeeded in
1508 by Richard Pynson, with a salary of £2, increased to £4 in 1515. Wynkyn
de Worde was printer to the Queen Mother Margaret in 1508 until her death
in 1509.
[1] Caxton, or his manuscript, mistakenly supposes that *Polycronicon* ends in
1357 (in fact it includes the Treaty of Bretigny, 1360), and that Trevisa completed his translation in the same year 1357.
[2] By the Carthusian monk Werner Rolewinck of Cologne, first printed in
Cologne in 1474. Possibly Caxton used his master Veldener's edition, Louvain,
29 December 1475, of which a copy in the British Library was bought on 5 June
1479 by Richard FitzJames, later Bishop of London.
[3] Identified by Blake as the *Historia aurea* of John of Tynemouth, monk of
St. Albans, ending in 1347, but continued, in a manuscript once at St. Albans but
now at Corpus Christi College, Cambridge, as far as 1377. If so, this may be
another of the many signs of Caxton's contact with St. Albans, where the
Benedictine monastery maintained close relations with Westminster Abbey.

edition of the story of King Arthur, so here he announced his intention to publish the *Golden Legend* as a sacred companion to the secular *Polycronicon*: 'Then since history is so precious and also profitable, I have delibered to write two books notable, retaining in them many notable histories, as the lives, miracles, passions and death of diverse holy saints, which shall be comprised in one of them which is named *Aurea Legenda*, that is the *Golden Legend*, and that other book is named *Polycronicon*.' Perhaps with a similar motive he mentioned that Trevisa had also translated the Bible and Bartholomaeus Anglicus. Trevisa's Bible translation is known only from this state-ment of Caxton's; but Caxton's *Golden Legend* is unique in containing an Old Testament abridgment which is of unknown origin, and it seems worth suggesting (though less likely than the alternative hypothesis which we shall examine in due course) that this may derive from Trevisa's lost work. It is still more probable that Caxton meant to print Trevisa's *Bartholomaeus*, for this work was at last produced by Wynkyn de Worde towards 1495 as one of a series of books which seem to be commissions or intentions left unfulfilled by Caxton's death in 1491, together with the famous verses which reveal that Caxton had learned his art by printing the Latin original at Cologne.

No doubt Caxton had already begun his translation of *Golden Legend*, probably soon after finishing *Godfrey of Boloyne* on 7 June 1481, which he lists as his last previous translation. He bemoaned 'the long time of translation', and indeed *Golden Legend*, which he completed on 20 November 1483 and printed in 1484, is by far the longest of his translations, containing over 600,000 words. The preoccupation of this tremendous task, along with the political uncer-tainties which followed Edward's death, explains his preference for second editions and other readymade texts until the end of 1483.

Three such books, all undated, comprising a third edition of *Cato*, a second of *Game of Chess*, and *Curia sapientiae*, can be assigned on typographical grounds to the period between *Chronicles II*, 8 October 1482, and his next dated book, *Pilgrimage of the Soul*, 6 June 1483. One such argument is the curious fact that in the seven books[1] produced in the period between Caxton's first use of printed signa-tures in *Chronicles I*, 10 June 1480, and the completion of *Chronicles II*, 8 October 1482, the leaf numbers within the quire are printed mostly or entirely in arabic numerals, whereas in these three undated books and in all Caxton's later work roman numerals are used.[2] *Cato III* and *Game of Chess II* must have been printed in close succession, for paper with a jug watermark is used in each, each is illustrated with woodcuts, and in each type 2* (which here makes its last appearance as a text type) is found in the same peculiar state, with more wrong-

[1] i.e. *Chronicles I, Mirror, Reynard, Tully, Godfrey, Polycronicon, Chronicles II.*
[2] With the isolated exception of Gower, *Confessio amantis*, 2 September 1483, in which arabic numerals are used in the first half of the book, and roman numerals in the rest.

fount majuscules from type 3 (including A, B, C, E, H, T) than in any other books. This contamination must have occurred during the printing of *Cato III*, in which type 3 is used in large quantities for the Latin text, so *Cato* evidently came first. *Cato III* is in folio, unlike the two previous editions which are in quarto, and must have been reprinted from *Cato II*, with which it shares various textual differences from *Cato I*. In the middle sheet of the first quire both the Latin and the English text are in type 2*. This not only confirms that Caxton by this period was printing 'by the forme', two pages at a time, but also shows that the compositor was setting from the middle of the quire outwards, a common process which was only made possible by previous accurate casting off. *Cato III* is adorned with two comical woodcuts of master and schoolboys previously used in *Mirror of the World*.[1] *Game of Chess II* has a vigorous and wellknown series of sixteen new cuts, evidently adapted (as were the great majority of fifteenth-century printed illustrations) from drawings or illuminations in a manuscript, and by a different and slightly more competent cutter. For obvious reasons Caxton omitted the dedication to the now unmentionable Clarence, and substituted the briefer autobiographical prologue which tells how he first read and translated the book 'at such time as I was resident at Bruges in the county of Flanders', and then 'did do set in emprint a certain number of them which anon were depeshed [dispatched] and sold'. Oddly enough, this is Caxton's only outright statement of the fact (indubitable on other grounds) that he printed at Bruges before coming to Westminster, for none of his Bruges books specifies the place of printing.

Curia sapientiae, or *Court of Sapience*, is a dream allegory in verse of a controversy between Mercy, Truth, Justice, and Peace, the daughters of a great king with a beloved son, followed by a tour through the Palace of Wisdom. The poem was later ascribed to Lydgate,[2] but is left anonymous by Caxton, and is rejected by modern Lydgate specialists on stylistic grounds, with the suggestion that the reigning monarch to whom the author addresses it was not Henry VI but Edward IV, and its date about 1475. The poem, though ostensibly concerned with the Kingdom of Heaven, is evidently a compliment to Edward, whose beloved son was the little Prince, and whose four daughters Elizabeth, Mary, Cecily, and Anne were growing in marriageable importance.[3] Perhaps the most likely time for Caxton's edition is just after Edward's death in April 1483, when, as we shall

[1] Blades argued that *Cato III* preceded *Mirror*, alleging that the two cuts are more appropriate to *Cato* and show increased damage in *Mirror*. In fact the cuts belong to the original *Mirror* series as found in the parent manuscripts, and their state is identically the same in both books.

[2] First by Stephen Hawes in his *Pastime of Pleasure* (1506), line 1357 in the EETS edition (1928), and so in Duff, STC, etc.

[3] Elizabeth (1466-1503) was betrothed to the five-year-old Dauphin of France as part of the Treaty of Picquigny in 1475. Mary (1467-82) died young, but Catherine (1479-1527) and the nun Bridget (1480-1517) arrived to make five.

see, he engaged in other propaganda for the royal princesses. However this may be, *Curia sapientiae* is the only undated book in type 4 with signatures in roman numerals; so it doubtless belongs to the brief period after the arabic-numeral series and before the introduction of type 4*—that is, between October 1482 and June 1483.

Caxton's fourth and last work printed entirely in type 3, a quarto Latin *Psalter*, perhaps appeared within the same uncrowded interlude of undated books. Its arabic-numeral signatures might tend to place it in 1482, but its paper points rather to 1483, for the intrusive watermark (in quire i only) of a double rose made in Normandy was found by Allan Stevenson to be in an earlier state than an occurrence of a twin mark in Normandy archives in 1484.[1] By a mishap just like the one in *Reynard* the compositor was obliged to insert a separate leaf containing two pages of omitted text.[2]

Caxton's other book in type 3 is a quarto Latin *Horae ad usum Sarum* or *Book of Hours* in the Sarum rite. These immensely popular books of prayers for the liturgical hours of the day, a lay equivalent of the priest's breviary, were printed in several hundreds of editions for the various national or local rites all over Europe from the mid 1480s, especially in Paris. Caxton, rather unexpectedly, was first in the field. He had already produced an octavo edition, for which the irregular line endings and the use of type 2 in its first state suggest a date towards 1477-8.[3] *Horae II*, of which a perfect copy would contain about 230 leaves, survives only in four mutilated leaves, taken from the *Boethius* binding discovered by Blades at St. Albans Grammar School, and even these are press waste discarded after a frightful error in printing.[4] Perhaps *Horae II* in type 3 was printed about the same time as *Psalter*; but more probably, as the line-endings are slightly irregular, it belongs with *Dicts II* to early 1480 or late 1479, soon after the type 3 *Advertisement* and *Ordinal*.

So Caxton arrived at the spring of 1483, in which the Shakespearean catastrophes that destroyed his patrons, and the long-awaited introduction of the new casting of his type 4, were for him events of equal or counterbalancing importance.

[1] But Stevenson suggests, very cogently, that the intrusive rose paper may mean that quire i is a cancel—that is, a later reprint to make up a miscalculation of the numbers required, a frequent mishap in Caxton's office. In this case, quire i could be assigned to 1483-4, and the rest of the book to 1482 where it more naturally belongs.

[2] In the penultimate quire, between leaves x7 and x8. The 'homoeoteleuton' word in this case is '*Per*', with which both the inserted leaf and the one before happen to end. The accident, once again, is a symptom that casting off had occurred.

[3] A 62-leaf fragment in the Pierpont Morgan Library is the earliest of Caxton's four known books (excluding Indulgences) on vellum, and is unique among English incunables in being finely illuminated (though only with floral borders and capitals, not with the scriptural scenes found in contemporary French and Flemish manuscript *Horae* and later in the Paris printed *Horae* of the 1490s).

[4] A 2-leaf fragment printed on the inner side only omits about 50 leaves of text between the lefthand and righthand page! Probably the book was printed on two presses, and a page of type from one was inadvertently imposed in a two-page forme beside a page from the other.

13

UNDER WHICH KING?

EDWARD IV, though ailing and 'grown fat in the loins', kept Christmas 1482 at Westminster with more than usual splendour, as if he might live for ever, wearing 'a great variety of costly garments', including a new line in furtrimmed robes which 'gave that prince a new and distinguished air to beholders, he being a person of most elegant appearance', says a chronicler. But he came home from Windsor after Easter with a cold caught fishing, and died quite unexpectedly on 9 April 1483, aged almost forty-one. They embalmed him and took him back to Windsor for burial in the still unfinished Garter Chapel of St. George, which he had begun in 1475 (so that St. George's Windsor celebrates its quincentenary towards the same time as Caxton's). The Prince of Wales, aged twelve, still at Ludlow with his uncle and guardian Rivers, was proclaimed King Edward V on 11 April.

To Caxton the new reign seemed to promise more patronage than ever, and to his friends at court more power, for the Woodvilles now had a child king of their own blood. But Gloucester, suddenly becoming a 'bottled spider', arrested Anthony Earl Rivers at Northampton on 30 April and seized custody of the new King, who wept at this ominous change of uncles. Hastings, having vowed to the dying Edward to forget his quarrel with the Woodvilles and to 'love my lord Dorset', used his control of the Council to betray them to Gloucester; he changed his mind late in May, when he realised that Gloucester was aiming for the crown, and began counterplotting with the Queen Mother his enemy; but he rather deserved his lightning decapitation on 13 June.[1] Bishop Russell, Rivers's friend and Caxton's patron, ratted when Gloucester made him Chancellor twice over;[2] he joined the wheyfaced deputation which ravished the little Duke of York from the Abbey Sanctuary to the Tower on 16 June (on the ingenious plea that as the child had done no wrong he was not entitled to sanctuary); he prepared a snide sermon for the opening of parliament, praising the 'firm islands' of the old nobility in preference to 'the unstable and wavering running water of any great Rivers'. The man he punned against had named Russell among his executors in his will, wore a pious hair shirt for his decapitation at Pontefract Castle on 25 June,[3] and left a sad little poem, beginning:

[1] To do him justice, Hastings perhaps never understood that he had doomed Rivers to death and not to mere captivity. He boasted that the transfer of power had been made 'with less blood than a cut finger'.

[2] First when the loyalist Rotherham was dismissed early in May, again on 27 June, the day after Gloucester was proclaimed King.

Somewhat musing
And more mourning
In remembering
The unsteadfastness . . .

and ending:

My life was lent
Me to one intent,
It is nigh spent,
Welcome, Fortune! . . .

Gloucester (much to the annoyance of his mother Cecily Duchess of York, in whose London home at Baynard Castle he was living then) revived the old scandal used by Warwick and Clarence in 1470 and by Clarence in 1477, that his brother Edward was a bastard, being the result of an adulterous affair between their mother and an archer named Blackburn in France in 1441.[1] The new king and the little Duke of York were bastards twice over, because Edward had wedded a certain Lady Eleanor Butler by promise before his marriage to Elizabeth Woodville, which was therefore void. Hired preachers spread this surprising news in the City on Sunday 22 June, a scratch Parliament hailed King Richard III on 26 June, the usurper was crowned in Westminster Abbey (a few yards from Caxton's Chapter House shop) on 6 July, and the little Princes were secretly murdered in the Tower, probably towards the end of that month.[2]

So Caxton lost his royal patrons and friends at court, Edward, Rivers, Hastings, the little Princes for whom he had produced those two bumper books for boys, *Jason* and *Godfrey of Boloyne*, all in four months, by sudden death, the headsman's axe, and the smotherer's pillow. Edward Prince of Wales would certainly have been a bountiful master as Edward V, for he took after his literary uncle Anthony Rivers. The French agent Mancini reported as the boy's most salient characteristic 'his special knowledge of literature, which enabled him to understand fully and to declaim most excellently from any work whether in verse or prose that came into his hands, unless it were from among the more abstruse authors'.

But one still powerful patron remained, closer in abode and more

---

[3] Old Sir Thomas Vaughan, the Prince's chamberlain, and Richard Grey, the Queen's son by her first marriage, who had ridden from London to warn him to arrive ahead of Gloucester, were arrested and beheaded with Rivers.

[1] Charles the Bold, in his pique after their quarrel in 1475, made a point of referring to his brother-in-law Edward IV as 'Blayborgne'!

[2] Not as early as 26 June as some have suggested, because it was then to Richard's advantage to pretend that he was deposing them for illegitimacy but sparing their lives from magnanimity; nor as late as August-September, when Buckingham's rebellion depended not merely on their rumoured death but on knowledge that Richard could no longer produce them alive; but in July, when the Queen Mother and other surviving Woodvilles were conspiring to restore them, thus enabling Richard to destroy them as rivals and to plead, if necessary, that they were executed as traitors by law.

in need of Caxton's services than before. Towards midnight on 30 April the Queen Mother Elizabeth Woodville heard the appalling news of her brother's arrest at Northampton early that morning, and Gloucester's possession of the Prince. She fled from Westminster Palace across the way to the Abbey Sanctuary, taking her younger son the Duke of York, her five daughters, and her dead husband's treasure, 'chests, coffers, packs, fardels, trusses, all on men's backs, some lading, some going, some discharging, some breaking down the walls to bring in the nearest way'. At dawn 'all Thames was full of the Duke of Gloucester's servants, watching that no man should go to Sanctuary, nor none could pass unsearched'. Abbot Eastney gave her lodging in his own house, and Prior Essex got into serious trouble for harbouring the goods of her son Dorset.[1] The Queen remained in the Abbey, a prisoner but invulnerable, seemingly powerless but in fact one of the most dangerous persons in England as a nucleus for opposition to the usurper. Her presence within the same walls is an unnoticed but crucial factor in Caxton's career throughout the reign of Richard III.

In the brief unreal eleven weeks' reign of Edward V Caxton produced one book, *Pilgrimage of the Soul*, 6 June 1483, an anonymous English prose version of a French poem *Pèlerinage de l'âme* written about 1330 by the Cistercian monk Guillaume de Deguilleville. This has often been ascribed to Lydgate, partly because Lydgate made a verse translation of a companion piece by Deguilleville, *Pilgrimage of the Life of Man*, and partly because Blades and others have asserted that the text includes portions of a genuine Lydgate poem, *Life of Our Lady*. This notion has an odd but natural explanation, which has not previously been noticed. A copy of *Pilgrimage of the Soul* which was described in the Harleian Library sale catalogue in 1743 and was thereafter lost until it came to the surface again in 1931, contained quire e of Caxton's edition of *Life of Our Lady* (which was printed a few months later in 1483), in place of its own rightful quire e. The cataloguer of the Harleian copy (now in Yale University Library) mistook the quire inserted from *Life of Our Lady* for an integral part of *Pilgrimage of the Soul*, and accordingly suggested that 'this offers some likelihood that the Translation was made by Lydgate'. The substitution perhaps occurred by accident in Caxton's workshop when the loose quires were gathered together for binding this copy; if so, the mistake was an easy one, for the first page of quire e happens to bear a superficial resemblance in both books, being in seven-line stanzas of rhyme royal. The nameless

---

[1] Thomas Grey Marquess of Dorset (1455-1501), the Queen's elder son by her first marriage and Hastings's chief rival, and her brother Lionel Woodville, Bishop of Salisbury (1446-84), went into sanctuary with her on 30 April. Dorset escaped early in May, and Lionel soon after 9 June. Both helped to organise Buckingham's rebellion in August-October, and then joined Henry Tudor in Brittany. Dorset had brought some of Edward's treasure from the Tower to the Abbey. There was a great deal of money around in the Abbey in those months, and it seems very likely that Caxton got some of it on account.

translator (who remarks: 'I the simple and insufficient translator of this little book have in diverse places added and withdrawn little what as me seemed needful')[1] has incorporated no fewer than fourteen passages in English verse, one by Hoccleve,[2] the rest unidentified and perhaps his own composition, but not in Lydgate's manner or prosody. So the usual ascription of the translation to Lydgate seems completely baseless and ought to be abandoned.[3]

Perhaps Caxton or his clients thought the text specially appropriate to the recent death of his profligate but pardonable King, for it tells of the prosecution of a newly departed soul by Satan before Reason, Truth, and Justice, who consider his case hopeless, and Mercy, who secures his speedy acquittal. The parallel with the heavenly princesses in *Court of Sapience* is evident. Caxton dated his edition, rather pointedly, 'at Westminster' in 'the first year of the reign of King Edward the Fifth'. His public knew very well that the Queen Mother was in sanctuary within the same precincts as the Printer, her son captive in the Tower and unlikely ever to reign in fact, and that both had been his patrons in better days; they would get the message. Perhaps they also noticed that in his next book, Mirk's *Festial*, 30 June 1483, he did what he had never done before in any dated work since he came to Westminster: he omitted the regnal year altogether, and so avoided mentioning the name of Richard III, who had deposed his nephew only four days before.

Pilgrimage of the Soul, 6 June 1483, was Caxton's last book entirely in the first state of type 4, which he had begun to use early in 1480. Mirk's *Festial*, 30 June 1483, which must have been printed immediately after, or concurrently, is his first in the new state, 4*. Type 4 had now served the normal lifespan of three or four years, and was due for replacement; so had types 2* and 3, which he now used only occasionally and sparsely as mere heading types. Another motive, no doubt, was his need for an increased supply of type to serve extra printing presses in the production of the long-planned *Golden Legend* and *Morte d'Arthur*, his longest and most ambitious books. The change of type cannot have had any connection with the change of monarchy; Caxton must have ordered type 4* from Veldener early in the year,[4] and well before Edward IV's death. In the new 4* the

---

[1] This remark (not included in Caxton's manuscript or not printed by him) occurs in the translator's colophon in British Library MS. Egerton 615.

[2] Book iv, ch. 20 of *Pilgrimage* consists of Hoccleve's 'Complaint of the Virgin before the Cross' with five additional stanzas.

[3] Equally unjustified is the assumption (made by Blades and repeated by all) that the English translator used the French prose version of Deguilleville's poem written for John Duke of Bedford when Regent of France (1422-35) by his chaplain Jean de Gallopes. I have found from comparison of the three texts that the English translation represents an independent adaptation of the verse original, distinct from Gallopes's; so it must have been made either directly from Deguilleville's verse, or from some other French prose rendering, but not from Gallopes's.

[4] Veldener must have been glad of Caxton's order, for he was temporarily at a loose end. After leaving Louvain in 1477 in disgust at the competition of

face-size of the type remains unchanged, but the body-height is enlarged from 95 mm. to 100 mm. per 20 lines, thus giving a perceptible and agreeable increase of space between the lines, at the expense of reducing the normal page contents from 40 lines to only 38, and a corresponding rise in paper requirement. Most letters remain identical in shape, but the two states can be readily distinguished by A, which in type 4 usually has a decorative hairline to the right which is normally absent in 4*, and by w, which in 4 generally has a looped head-flourish and leans leftward, whereas in 4* a vertical unlooped w is normal.

John Mirk was Prior early in the fifteenth century of the great Abbey of Augustinian Canons[1] at Lilleshall in Shropshire. His *Festial* or *Liber Festivalis* is a collection of racy anecdotal English sermons for parish priests and their congregations, arranged in order of the feast days of the liturgical year (hence its title). Caxton's manuscript, besides differing throughout in wording, was less complete in contents than the one used in the Oxford edition of 14 August 1486, which was reprinted in England in the 1490s in a total of eight editions including one by Caxton himself. Among the omissions were the author's preface and colophon, so that Caxton's first edition has the peculiarity of leaving the book not only anonymous but completely without a title.

As a companion piece Caxton printed an undated edition of *Quattuor sermones*, a shorter collection of four sermons in English on the basic doctrines of the Church, including the Lord's Prayer, the Creed, the Ten Commandments, the Seven Deadly Sins, and the rules of confession. This became the inseparable consort of *Festial*, and the two works are generally found bound together both in Caxton's first edition and in the later editions of the century.[2] One of the four new Caxtons of this century was discovered by Christopher Webb of the National Central Library, when he noticed that different bibliographers gave conflicting descriptions of *Quattuor sermones*, and found that four of the nine known copies belong to an entirely different edition from the one generally known. Webb's new edition turns out to be the true first edition produced in 1483, for it contains no typographically late features, while the other must have been

---

Johannes de Westfalia he had printed at Utrecht in 1478-81, but moved on again, no doubt owing to the war of independence declared on 7 August 1481 by the anti-Burgundians in Utrecht against their Bishop David (Philip the Good's bastard) and Holland. When he cast type 4* for Caxton he was at Culemborg (in Guelderland just south of the Utrecht frontier), where he produced his first book on 6 March 1483. He returned to Louvain in 1484.

[1] Mirk calls himself '*canonicus regularis*', i.e. Augustinian Canon; but Blades, not knowing the meaning of *regularis* in this context, called him merely 'canon of Lilleshall', and was copied by Duff, Aurner, and others.

[2] Perhaps it was Caxton who first began the fashion of uniting the two works, for no manuscript of *Quattuor sermones* seems to have survived, whether separate or together. The 1486 Oxford *Festial* is exceptional in having no companion edition of *Quattuor sermones*.

printed a year later, in 1484, as it uses printed initials and paragraph marks which first occur in that year.[1] Evidently Caxton was obliged to reprint *Quattuor sermones* to meet an unexpected demand from customers who wanted both works bound together.

On 2 September 1483 Caxton completed an edition of *Confessio amantis*, a collection of love tales in verse by Chaucer's contemporary John Gower. This time (and in other dated books during Richard's reign), having made his point by omitting it in *Festial*, he gave the regnal year of the usurper, 'the first year of the reign of King Richard the Third'. Caxton's text combines certain characteristics of all three families of the manuscript tradition, and it has been supposed that he intentionally printed from three different manuscripts in order to present as complete a version as possible; but Blake's suggestion, that he used a now lost manuscript in which the three recensions had already been conflated, seems a much more likely explanation. So Caxton had now printed the major work of the third in his favourite trio of Chaucer, Lydgate, and Gower. He had quoted *Confessio amantis* (without acknowledgment) several times in his *Ovid* translation. Very likely Caxton had been familiar with Gower's book all his life; but possibly this clue implies that his intention to print it came from a rereading during his work on *Ovid* in 1479-80.

Confessio amantis is one of Caxton's longest texts, and forms a first instalment in the programme of large texts which he had planned in the year before. This and *Golden Legend* are his only books on large paper. In *Confessio amantis* he used the size then called 'bastard' (being halfway between the normal sized 'median' and the extra-large 'royal' which he was to use in *Golden Legend*), which gave him space for 46 lines of type 4 or 44 of 4* to a page; and Gower's short octosyllabic lines enabled him to save paper by printing for the second time (the first was in *Doctrine to learn French and English*) in two columns.[2]

His next dated work—in a period to which, as we shall see, a rather large number of undated works must also be assigned—was *Book of the Knight of the Tower*, 31 January 1484, translated by Caxton himself from a collection of moral or immoral tales made for the instruction of his daughters by Geoffroy de la Tour Landry in 1371, another Franco-Burgundian classic. The Knight's tower still exists (as Mrs. M. Y. Offord describes in her exemplary Early English Text Society edition of 1971) in the small town in Maine-et-Loire called La Tourlandry after it. His stories are sometimes disconcertingly ribald, or bawdy, or cruel; disobedient wives are beaten with staves, the ropemaker's wife has her legs broken with a pestle for

[1] The 1483 edition, first identified by Webb, is found in the John Rylands, Lambeth, St. John's College Oxford, and Vienna copies. The second edition is found in the British Library, Bodleian (two copies), and the Saint Andrews University and Huntington libraries.

[2] Mr. Nicolas Barker has pointed out to me that *Doctrine* also (which achieves a page-height of 42 lines in type 4 instead of the normal 40 or fewer) is probably on 'bastard' size paper.

el to haue torned her/But he myght not/ And when he sawe
hat she was ferme and constaunt/he lofte her/And after sayd/
nd tolde to many other the constannce and stedfastnes of her ,
herof he moche preysed and honoured her the more/And ther-
re here is a good Ensample/how me ought not to goo to fooly
elgremages for no foolysshe playsaunces/ But only for the dy-
yne seruyse and for the loue of god / And how good it is for to
rape/and to doo saye masses for the soules of fader/ moder and
ther frendes/ For in lyke wyse they praye and empetre grace for
em that ben alyue that remembre them/and doo good for them
s ye haue herd / And also it is good to gyue almesse for gods
ke/for the almesses geten grace of god to them/that gyue them
yke as ye haue herd/And nowe I shalle telle yow another en-
mple that happed in a Chirche/whiche was called oure lady of
realem

Of the man and woman that made fornycacion within the Chirche/ Capitulo   xxx

T befelle in the same chirche vpon the vygyl or euen of
oure lady that one named Perrot Lenard/whiche was
sergeaunt of the saide chirche that same yere laye with a
woman vnder an awter/in whiche place this myracle befelle/
they were ioyned to geder as a dogge is to a bytche/And in this
manere they were founden & taken / & so ioyned & knytte to geder
they were all the hole day / in so moche that they of the chirch &
of the Countrey had leyser ynough to see & behold them. For they
mought not departe one fro another/wherfore a processyon was ma
e for to pray god for them/And soo aboute the euenyng of the
daye they were losed and departed that one fro that other /
Neuertheles nedeful and right it was that the Chirche sholde be
newe halowed , And that the said perrot for his penaunce sholde
goo al about the Chirche al naked on thre sondayes/betyng hym
elf/wepyng and tellyng his defaute and synne , And therfore
ere is to euery man a good ensample/how that he sholde hold hym
clenly and honestly in holy chirche. And yet shall I telle yow an
other ensample vpon the same matere, whiche byfelle in the par-
tyes of pycardy, whiche is not past thre yere

D ij

Caxton's Type 4, from Knight of the Tower, 1484.

127

adultery with the prior, not that that stops her. But the Knight himself is hardly to blame, for he writes within the *genre* of the late medieval sermon collection or devotional storybook, on the principle that a sensational tale makes its pious moral still more improving; and he expresses with touching simplicity his affection for his wife Jeanne and 'my well beloved daughters whom I see so little'.[1] Besides compiling familiar tales from other collections, the Knight tells many autobiographical anecdotes of his own youth, his fellow warriors, his relatives, or local scandals, such as the sad case of Perrot Lenard, sergeant of the nearby town of Candé, who remained miraculously and inextricably joined to a woman with whom he lay under the altar in a church in the Knight's own domain, 'in so much that they of the church and of the country had leisure enough to see and behold them'. Exactly the same mishap occurred to a monk named Pigière in Poitou, and the moral is that one ought not to look at a woman in church except 'by thought and way of marriage'. But perhaps the Knight's daughters were still more edified by the Thurberish cautionary tale of the talking magpie that informed its mistress's husband when she and her chamberer [maidservant] ate his favourite eel. 'And in the house therefore was great sorrow and noise. But when the lord was gone out the lady and the chamberer came to the pye and plucked off all the feathers of his head, saying: 'Thou has discovered us of the eel,' and thus was the poor pye plumed and lost the feathers of his head. But from then forth on if any man came into that house that was bald or pilled [tonsured] or had an high forehead, the pye would say to them: 'Ye have told my lord of the eel.'

Caxton had finished his translation seven months before, on 1 June 1483, just before completing the printing of *Pilgrimage of the Soul* on 6 June. The book, he tells us, 'is comen into my hands by the request and desire of a noble lady which hath brought forth many noble and fair daughters which be virtuously nourished and learned'. Blake's suggestion that this lady was none other than the Queen Mother Elizabeth Woodville,[2] and that the daughters were hers, is certainly justified. However, the political implications of the Queen's commission to Caxton can be taken still further. At this very time, as we have seen, she and her five daughters (Elizabeth aged seventeen, Cecily aged fourteen, Anne aged seven, and the infants Catherine and Bridget) were captives in the Abbey Sanctuary, and Caxton's

[1] In this charming expression 'little' means 'young', not 'seldom'. Mrs. Offord quotes the complaint of Fitzherbert in his *Book of Husbandry* (1534), that the Knight of the Tower 'hath made both the men and the women to know more vices, subtlety and craft than ever they should have known if the book had not been made'.

[2] As Blake also points out, Caxton invites readers of his book to praise 'the lady that caused me to translate it, and to pray for her long life and welfare, and when God will call her from this transitory life that she may reign in heaven sempiternally'. The surprising word *reign* can only mean that his patroness was a queen.

near neighbours. These girls were now of supreme importance to the Queen and the Woodville family in their struggle against Gloucester, as heiresses of the blood royal and as political counters. Caxton's commission was no doubt arranged immediately after their flight to the Abbey on 30 April, and the completion of the translation inside a month was by no means beyond his powers when faced with a rush job. *Knight of the Tower* was no doubt intended for press immediately after *Pilgrimage of the Soul*, but was rendered inopportune by the lightning events of that June, so that the innocuous *Festial* had to be substituted. Richard was caught in his own mousetrap. By destroying the normal succession to the throne he had raised two distant outsiders, his own false ally Buckingham and Henry Tudor, to the position of dangerous claimants.[1] By liquidating the Woodville and Hastings parties he left himself with no party at all except his own creatures, 'the Cat, the Rat, and Lovel our Dog'. The nation was as outraged by his homicidal methods as modern historians ought to be. He had also enhanced the political significance of the two elder princesses, Elizabeth and Cecily, who would now be offered in marriage to the future king of England. Late in July 1483, after Richard's departure on his face-showing tour of the midland counties, it became known that the Queen and her advisers in Westminster Sanctuary were planning to smuggle the two girls 'in disguise to the parts beyond the sea, so that the kingdom might some day fall again into the hands of the rightful heirs'. Then 'the noble church of the monks at Westminster assumed the appearance of a castle and fortress,' says the chronicler, for Richard set his trusty captain John Nesfeld to guard the Sanctuary, 'so that not one of the persons shut up could go forth, and no one could enter without his permission'. Henry's mother Margaret Beaufort sent her Welsh physician Lewis and her priest Christopher Urswick (whom even Nesfeld could not stop). By September the Queen had promised to marry Princess Elizabeth (or, failing her, Cecily) to Henry Tudor if he became king. But the instant collapse of Buckingham's rebellion and Henry's support left the unhappy Queen resourceless. Henry swore in Rennes Cathedral on Christmas Day 1483, before his followers and her escaped son Dorset, to marry Elizabeth, too late. In January 1484 Richard countered this renewed threat to his throne by sending 'grave persons' to propose a deal with the Queen. Suddenly the printing of *Knight of the Tower*, which at any time in the previous seven months would only have warned the usurper and exposed her

[1] Henry claimed as great-great-grandson through his mother Margaret Beaufort of Edward III's son John of Gaunt. Buckingham's claim was almost as strong, as he not only had a Beaufort mother but was a great-great-grandson of John of Gaunt's younger brother Thomas of Woodstock; but he was not marriageable, as Edward IV had cleverly married him in boyhood, about 1466, to the Queen's sister Catharine Woodville. They joined forces because each intended to do away with the other after overthrowing Richard. Once again, their actions and the Queen's simply do not make sense except in the light of certain knowledge of the little Princes' death towards the end of July 1483.

plans, became a means of courting his offers and raising the price of her daughters; and this, no doubt, is precisely why the book appeared at last on 31 January 1484.

The new state of type 4 must have arrived in June 1483 by instalment only, for Caxton continued to use the old state concurrently. The first four-fifths of *Confessio amantis*, 2 September 1483,[1] is printed in type 4, the rest in 4*. The change-over occurred in quire z, where an extraordinary state of affairs is revealed. The compositor evidently began at the middle sheet of the quire (a process which was quite normal, as we have seen, thanks to casting off); but after he had set all these four pages in type 4, with the exception of the second column of the first page, his colleague on the other press[2] took over and set this remaining column in 4*, continuing thereafter with the rest of the quire and the book. This accident probably means no more than that the first compositor went out for a drink before his stint of one sheet a day was finished, or that his colleague was delayed from relieving him at the expected time. But ordinarily such a vagary would occur undetectably, for Caxton's use of two distinguishable states of type in the same book is exceptional, both in his work and in the whole field of fifteenth-century printing. *Knight of the Tower*, 31 January 1484, was shared more evenly between two presses, with the first six quires in type 4 and the last five in 4*.

Only these two books were printed in both states of the type, and no undated books in the earlier state can be assigned to the period after the first introduction of 4* in *Festial*, 30 June 1483. In Caxton's next dated book after *Knight of the Tower*, Aesop's *Fables*, 26 March 1484, he used printed initials and paragraph marks which were made specially for *Aesop*, and were in general use thereafter. The use of both states in equal portions in *Knight of the Tower* can easily be explained, for the book was a rush job for the Queen and required all Caxton's resources. But the use of type 4 for a greater part of *Confessio amantis* must surely mean that 4* was being used for other work in July–August 1483. Similarly the absence of dated work, excepting *Knight of the Tower*, in the six months between *Confessio amantis* and *Aesop*, suggests that 4* was then being used for undated books. In fact a substantial group exists of dateless books which fulfil the required conditions, since from the use of 4* they must be later than June 1483, and from the absence of printed initials and paragraph marks they must be earlier than March 1484. Their total quantity of printed matter, a little under 600 leaves, seems just right for filling the available gap. In my view these books, which by Duff were more loosely dated as '1483–5', can be assigned with confidence

---

[1] Perhaps it is no coincidence that even this apparently harmless classic was printed by Caxton, in that month of Lancastrian pretenders, in the version dedicated by Gower to Henry of Lancaster the future King Henry IV, replacing the earlier dedication to Richard II.

[2] Or rather, on one of the other presses. At this time Caxton was probably using four presses, two with type 4 and two with 4*.

to the period from July to December 1483. The books are Chartier's *Curial*, Chaucer's *Canterbury Tales II*, *Book of Fame*, and *Troilus and Criseyde*, Lydgate's *Life of Our Lady*, and *Sex epistolae*. Caxton must surely have had some reason for leaving such a homogenous group undated; and perhaps his motive may be found in his continuing reluctance to give the regnal year of the usurping Richard III.

Perhaps *Curial*, for political reasons, came first. This brief folio pamphlet of six leaves, which can only have taken a day or two to print, is 'the copy of a letter which Master Alain Chartier wrote to his brother which desired to come dwell in court, in which he rehearseth many miseries and wretchednesses therein used, for to advise him not to enter into it lest he after repent . . . which copy was delivered to me by a noble and virtuous Earl, at whose instance and request I have reduced it into English', says Caxton. Blades and Blake are surely right in suggesting that the Earl was none other than Anthony Rivers so lately beheaded,[1] and everyone would understand that Caxton's transparent but prudently non-actionable allusion bore witness on the Queen's behalf against the usurper and murderer. Chartier's savage satire on life at court—where 'if he study for to find friendship he shall never can trot so much through the halls of the great that he shall find her'—was well suited to convey the sentiments in life and the message from beyond the grave of the romantic, literary, embittered Earl, who in *Cordiale* had denounced 'the abominable and damnable sins which commonly be used nowadays, as pride, perjury, terrible swearing, theft, murder, and many other'.[2] Caxton made one of his few ventures into verse (though without attempting to reproduce the complex rhyme scheme of the original) by including a humorous, mock-proverbial ballade by Chartier, beginning:

> *There is no danger but of a villain*
> *Nor pride but of a poor man enriched,*

which he emphatically signed 'Caxton' at the end.

Likewise he printed his name 'Caxton' in the margin in *Book of Fame* (as he called the poem nowadays oftener known as *House of Fame*), to mark the point where Chaucer's never-completed text broke off in Caxton's defective manuscript[3] and his own makeshift

---

[1] Not on 13 June, as Blake says, but on 26 June 1483; it was Hastings who lost his head on the 13th. *Curial*, I think, came before the end of July, when the Queen had her sons' as well as her brother's death to lament.

[2] The French *Curial* is a close translation of the Latin *De miseria curiali*, the priority of which is established by the presence of misunderstandings of the Latin in the French. Scholars now agree that the translation is by Chartier himself.

[3] In the completest manuscripts Chaucer's poem runs to sixty-four further lines, but still remains unfinished. Caxton's closing couplet is not too bad:

> *Thus in dreaming and in game*
> *Endeth this little Book of Fame.*

twelve-line conclusion began. 'I find no more of this work toforesaid,' he remarks, 'for as far as I can understand this noble man Geoffrey Chaucer finished at the said conclusion of the meeting of Lesing and Soothsay'. Caxton proceeded with his wellknown and rather under-rated praise of Chaucer: 'In all his works he excelleth in mine opinion all other writers in our English, for he writeth no void words but all his matter is full of high and quick sentence. To whom ought to be given laud and praising for his noble making and writing, for of him all other have borrowed since and taken in all their well saying and writing.'

Another piece of Caxton's verse appears as an epilogue in his edition of Lydgate's *Life of Our Lady*, a 6000-line poem on the life of the Virgin from her marriage to the child Christ's Presentation in the Temple, with many didactic digressions:

> *Go, little book, and submit thee*
> *Unto all them that shall thee read*
> *Or hear, praying them for charity*
> *To pardon me of the rudehead*
> *Of mine emprinting not taking heed . . .*

The second and third sheets of quire a are also found (not in any of the seven known complete copies, but only as fragments of press waste used in bindings) in a variant setting which has sometimes been explained as part of a lost second edition; but these two variant sheets are probably merely reprints necessitated by miscalculation of the numbers required, as so often occurs in Caxton's output. Since no one has yet compared every existing copy of every Caxton edition, more such variants no doubt await discovery. A hitherto unnoticed example occurs in *Troilus and Criseyde*, in which I have found different settings of the outer sheet of quire m in the two British Library copies.[1] For *Troilus* Caxton used a manuscript, now lost, of Chaucer's revised version, without noticing that some leaves were missing, and others misplaced; but even so his edition remains an important source for this text. The book is equalled in Caxton's output only by *Canterbury Tales I* and *Morte d'Arthur* for the union of visual splendour with the supernatural power of great literature in a first printing.

Canterbury Tales II, in contrast, takes an air of jollity rather than magnificence from its twenty-six cheerful woodcuts of the Pilgrims on horseback, or sitting at supper with the Host at a round table. These were evidently executed by the *Game of Chess II* cutter working as usual from illustrations in a manuscript, perhaps (but not necessarily) the one from which Caxton printed the text. Characterisation

[1] Distinguishable by the usual abundance of compositorial differences in spelling; for example, in the first line of m1 recto one copy (shelfmark G.11589) reads 'herte', while the other (C.11.c.10) reads 'hert/'. The latter copy contains the signature ('Iane Dudley') of the hapless Lady Jane Grey (1537-54); did she perhaps inherit it from her greatgrandfather Dorset, who was in the Abbey Sanctuary with his mother the Queen a few months before the book was printed?

Good wyf ther was of beside bathe
And she was somdeel deef & that was scathe
Of cloth makynge had she suche an haunt
She passyd them of ypre and of gaunt
In al the parisshe wyf was ther non
That to the offrynge before her sholde goon
And yf ther dyd certayn wroth was she
Than was she oute of al charyte
Her kercheups ful fyn were of grounde
I durste swere they weyed thre pounde
That on a sonday were on hyr hed
Hyr hosyn were of fyne scarlet reed
Ful streyte I tyed and shoos ful moyst and newe
Bolde was her face fayr and rede of hewe
She was a worthy woman al hyr lyue
Husbondys at the chyrche dore hadde she fyue
Withoute other companye in youthe
But her of nedyth not to speke as nowthe

4 Caxton's Type 4*, from Canterbury Tales II, 1483.

is uneven. The Miller, who would have been profanely aggrieved to see it, is portrayed as a slim boy with a flute, but the Wife of Bath, riding sidesaddle with a peekaboo hat and a naughty look, is herself for ever.[1] For his text Caxton used a manuscript supplied by a dissatisfied purchaser of his first edition. This, which he had printed unwittingly from an 'incorrect book brought to me six year past', was 'sold to many and diverse gentlemen, of whom one gentleman came to me and said that this book was not according in many places unto the book that Geoffrey Chaucer had made, to whom I answered that I had made it according to my copy and by me was nothing added nor minished. Then he said that he knew a book which his father had and much loved that was very true and according to his own first book by him made, and said more, if I would emprint it again he would get me the same book for a copy, howbeit he wist well that his father would not gladly depart from it. To whom I said in case that he could get me such a book true and correct yet I would once endeavour me to emprint it again for to satisfy the author, whereas tofore by ignorance I erred in hurting and defaming his book in diverse places in setting in some things that he never said nor made and leaving out many things which be requisite to be set in it. And thus we fell at accord and he full gently got of his father the said book and delivered it to me, by which I have corrected my book as hereafter all along by the aid of Almighty God shall follow.'

It is generally agreed that Caxton, as his own words ('by which I have corrected my book') seem to imply, did not print directly from his new manuscript, but from a copy of his first edition annotated with the necessary alterations in order of tales, wording, deletions, and insertions. This common practice was repeated by Pynson towards 1492 and Wynkyn de Worde in 1498, when they printed third and fourth editions of *Canterbury Tales* from their own independently marked copies of Caxton's second edition. Similarly in 1496 De Worde printed his *Book of Hawking, Hunting, and Blasing of Arms* from a heavily annotated copy of the St. Albans Schoolmaster Printer's edition of 1486.[2] The result is a unique hybrid between family 'b' of the manuscript tradition, from which Caxton's first edition derives, and family 'a', to which his new manuscript apparently belonged. Caxton's motives in this painstaking but patchy textual revision were no doubt mixed. He was glad, as always, to be on amicable terms with 'a gentleman', liked to feel he was doing his favourite poet a good turn, took a professional pride in the editorial tradition, rare among printers, which he had acquired from Colard Mansion and his fellows, and saw the story made a good selling point. But modern efforts to prove or disprove that in *Canterbury Tales II* Caxton shared the aims and abilities of the modern textual editor are anachronistic and inapplicable.

---

[1] This did not prevent Blades from reproducing the blameless Prioress in habit and rosary with the caption 'the Wife of Bath'.

[2] Discovered by Alan G. Thomas in 1961 and now in the British Library.

Unfortunately Caxton's mention of 'six years past' is of little use for dating purposes. It indicates only the interval of time between his receipt of the manuscript for edition I and the writing of his prologue for edition II, and does not relate directly, as many have assumed, to the printing of either.[1] The opening and closing dates remain unascertained or inferential. Perhaps the six years runs from the autumn of 1476, when as Stevenson suggests Caxton began edition I but postponed its completion till 1478, to autumn 1482, when he planned the programme of larger-scale works of which *Canterbury Tales II* and *Confessio amantis* were first to appear. The actual printing of Edition II probably occurred in July-August 1483 concurrently with *Confessio amantis*, and the consequent pressure on his type resources accounts for the use of the old type 4 in the latter work. *Book of Fame*, in which Caxton's eulogy of Chaucer includes abbreviated allusions to his *Canterbury Tales II* prologue, perhaps came next, and then *Troilus and Criseyde*, in which Caxton abstains from comment, having said his say already.

To the same autumn of 1483 belongs a short Latin quarto, *Sex epistolae quam elegantissimae*, or, as the colophon may be translated, '*Six letters of the extremest elegance, three from the Supreme Pontiff Sixtus IV and the Holy College of Cardinals to the Most Illustrious Doge of Venice Giovanni Mocenigo, and the same number from the Doge himself to the same Pontiff and Cardinals, on the War of Ferrara, printed by William Caxton and diligently emended by Petrus Carmelianus Poet Laureate, at Westminster*'. The editor added a Latin quatrain of his own composition:

> *Seeker of eloquence, purchase these six letters,*
> *Even Cicero could hardly write their betters . . .*

Carmelianus (1451-1527) was one of the itinerant Italian humanist busybees, like Traversagni and Surigonus, who buzzed all over Europe looking for honey, and found the flowers sweetest in England. He had arrived early in 1482 from France and Brittany on his way to Switzerland and Germany, but decided to stay, 'captivated as I was by the amenity and dulcitude of this land', found a temporary civil service job with the Keeper of the Rolls, and dedicated a Latin poem on spring to the Prince of Wales on 7 April 1482. Next, switching his bets, he addressed another on the martyred St. Catherine of Egypt, first to Richard's Chancellor Russell, and then to the new Constable of the Tower and accessory in the murder of the Princes, Sir Robert Brackenbury, with a panegyric on Richard as a model of all virtues. He struck lucky at last with a poem on the birth of Prince Arthur in 1486, and Henry VII made him his Latin secretary and loaded him with church benefices.

Did Carmelianus, hard up though he then was, commission Caxton

---

[1] Hence 1484, which typographically is a year too late, has often been assumed as the printing date for *Canterbury Tales II*, calculating from 1478, the probable year for the completion of *Canterbury Tales I*.

to print *Sex epistolae* as part of his ultimately successful campaign to get into the home diplomatic service of England? Or did Caxton pay him to see it through the press, a possibility which is strengthened by Carmelianus's move to Oxford in 1484-5, where he worked in the same way as a humanist editor for the printer Theodoric Rood?[1] Neither alternative can be the whole truth, for the text shows that *Sex epistolae* was not published solely for the 'extremest elegance' of its Latinity, or the self-advertisement of Carmelianus. In December 1482 Sixtus IV made a separate peace in his war against Ferrara and Naples, and on 30 April 1483 declared war on his own allies the Venetians when they decided to fight on. The Venetians appealed for support to France, the Emperor Frederick, even the Turk. *Sex epistolae*, although its politically sensitive nature is tactfully disguised by Carmelianus's title, is in fact a Venetian White Paper, publishing the exchange of letters between Venice and Sixtus from 12 December 1482 to 14 February 1483 with a preface arguing the justice of the Venetian cause, and suggests that the Venetians also hoped to influence England. Caxton's edition must have been paid for with Venetian money. Perhaps Carmelianus, who came from Brescia on the Venetian mainland, acted as middleman and subtracted his own fee; and perhaps part of the edition was used for distribution on the Continent in the Venetian interest.[2] Blades, who rarely erred in typographical matters, asserted that *Sex epistolae* is printed (like *Confessio amantis* and *Knight of the Tower*) in types 4 and 4* together, and was believed on trust by Proctor and Duff. In fact only type 4* is used, with a few lines of display in type 3.

The completion of his *Golden Legend* translation on 20 November set Caxton free to translate a new version of Cato, *Disticha*, which he finished on 23 December 1483 and probably printed in February 1484, immediately after the completion of *Knight of the Tower* on

---

[1] Carmelianus wrote prefatory verses for Rood's two undated editions of *Compendium totius grammaticae* by John Anwykyll, grammar master of Magdalen College School, towards 1484, and edited Phalaris, *Epistolae*, 1485, for Rood and the Oxford stationer Thomas Hunte. Weiss has been misled by Duff's merely notional dates for these books into dating the Anwykyll as 1483 and *Sex epistolae* as 1485, thus reversing their true order.

[2] The only surviving copy of *Sex epistolae*, now in the British Library, was discovered in 1874 at Halberstadt in Saxony in a seventeenth-century tract volume. Similarly Traversagni's *Nova Rhetorica* is extant at Turin, Savona, and Uppsala as well as at Corpus Christi College Cambridge, and *Commemoratio lamentationis BVM* in a unique copy at Ghent. The unique *Infantia Salvatoris* at Göttingen (now in Pierpont Morgan Library, New York), however, does not indicate export, as it was acquired at the Harleian Library sale in 1746. Venice secured favourable terms with bribery and French support at the Treaty of Bagnolo, 4 August 1484. Richard was no doubt committed against Venice, for he was about to ask Sixtus for a cardinalate for his Vatican envoy John Shirwood Bishop of Durham, France showed signs of supporting Henry Tudor, and the Parliament of January 1484 demanded restrictions against Italian merchants. Sixtus died of shock (12 August 1484) on hearing of the peace, and Richard transferred his new envoys Thomas Langton and Caxton's acquaintance John Kendale to the new Pope Innocent VIII.

31 January, and before beginning the *Aesop* of 26 March 1484. The last quire of this *Cato IV* contains the printed paragraph marks which were ordered for use in *Aesop*, as also does the last quire of *Quattuor sermones II*; so it seems that both these books were printed concurrently, and finished just at the moment when the new type material became available.[1] For this fourth edition, *Cato IV*, Caxton translated from a different text, in which the Latin original was accompanied by a French prose rendering with a commentary including *exempla* or moral tales.[2] But he took the opportunity to mention in his prologue the version which he had printed three times previously, in the English verse translation made about 1440 by Benedict Burgh for Sir William Bourchier the child son and heir of Henry, 1st Earl of Essex. All three, as it so happened, had died only a few months before. Old Essex, then nearly eighty, was an uncle of Edward IV (having married Edward's father's sister Isabel of York) and a tremendous Yorkist stalwart, and his son Sir William had married a Woodville, the Queen's sister Anne.[3] Caxton's in memoriam would be understood as discreet propaganda for the old regime, and was perhaps paid for by the Queen.

But his dedication of the whole book is unexpected. 'Which I present unto the City of London,' he says, 'I William Caxton citizen and conjury [sworn member] of the same and of the fraternity and fellowship of the Mercery owe of right my service and good will and of very duty am bounden naturally to assist, aid and counsel as far forth as I can to my power, as to my mother of whom I have received my nurture and living.' Yet his attitude towards the City, strangely enough, is much more critical than ingratiating. 'And shall pray for the good prosperity and policy of the same during my life, for as me seemeth it is of great need, by cause I have known it in my young age much more wealthy, prosperous and richer than it is at this day, and the cause is that there is almost none that intendeth to the common weal but only every man for his singular profit.'[4] His *Cato*, 'as in my judgement it is the best book to be taught to young children in school', will at least help to improve their awful offspring. 'By cause I see that the children that be born within the said City increase and profit not like their fathers and olders, but for the most part after that they be come to their perfect years of discretion and ripeness of

[1] The preliminary quire of *Cato IV*, including the table of contents, which of course was printed last of all, contains a single paragraph mark (except in a single variant copy discovered by Curt Bühler in the Pierpont Morgan Library).

[2] First printed about 1480 at Lyons by the anonymous Printer of the *Abusé en cour*; but no doubt Caxton as usual used a manuscript copy. He was again following in the steps of Colard Mansion, who had printed a different Latin-French version towards 1477.

[3] Essex died on 4 April 1483, his son on a date unknown not long before and Burgh on 13 July. As we have seen, Essex and Burgh both appear in the Little Wratting Caxton Deeds, and Essex's large-scale dealings in the cloth trade suggest a possible connection with Caxton the Printer.

[4] An echo of Caxton's outburst against English lawyers in *Game of Chess I*: 'for they intend to their singular weal and profit and not to the common'.

age, how well that their fathers have left to them great quantity of goods, yet scarcely among ten two thrive. I have seen and known in other lands in diverse cities that of one name and lineage successively have endured prosperously many heirs, yea a five or six hundred year, and some a thousand, and in this noble city of London it can unnethe [hardly] continue unto the third heir or scarcely to the second. O blessed Lord, when I remember this I am all abashed, I cannot judge the cause. But fairer nor wiser nor better bespoken children in their youth be nowhere than there be in London, but at their full riping there is no kernel nor good corn founden but chaff for the most part. I wot well,' Caxton hastily adds, feeling he may have gone too far, 'there be many noble and wise, and prove well and be better and richer than ever were their fathers.'

No doubt Caxton hoped to sell his book in the City, and perhaps, as Blake suggests, *Cato IV* implies a reversion to his merchant public to replace his Woodville patrons. Yet Caxton's prologue is a positive attack on the contemporary City, or at least on the new generation of merchants. Last June the Mayor's brother Friar Shaa had preached Richard's right to the throne, and the Mayor and aldermen had acclaimed him King, just as they had acclaimed Edward, Warwick, Henry, Edward again, and the little Prince. On 25 November 1483 they escorted Richard's triumphal return from the destruction of Buckingham's rebellion; at Christmas they lent him large sums of money on the royal treasures which had belonged to Edward. Caxton's blame for the decadent present and praise of the prosperous past would lose him more readers than it gained, for the timeserving newcomers who supported the usurper would know that the mercer printer appealed against themselves to the old Yorkist diehards of his own generation.

Aesop's *Fables*, 26 March 1484, was adorned not only with Caxton's new printed initials but with a sequence of 186 absurd and delightful animal woodcuts, evidently by the *Game of Chess II* cutter.[1] Here we have the only certain instance, among the many which have been claimed, of Caxton's use of a printed book for his model. His illustrations are crude imitations of those in the Lyons edition printed by Nicolaus Philippi and Marcus Reinhart, 26 August 1480. These in turn had been copied from the cuts in the edition in Latin, with German translation by Heinrich Steinhöwel, complete with the still more fabulous life of Aesop by the thirteenth-century Byzantine monk Planudes, various additional fables, and a choice of improper anecdotes from Poggio's *Facetiae*, printed about 1477 by Johann Zainer at Ulm. Caxton translated from the French

---

[1] Two more still less competent workmen were brought in towards the end to save time, a not uncommon occurrence in fifteenth-century book illustration when the press outpaced the woodcutter; but these were apparently responsible only for the actual cutting of the blocks from the first man's drawing, as the style of design remains little changed. The reality of the third man seems rather doubtful; he may be only the second man cutting one or two items in special haste.

version of Steinhöwel's collection by Julien Macho, an Augustinian monk of Lyons, as printed by Philippi and Reinhart, but with four more tales from Poggio, which Caxton probably found in a manuscript.[1] By Caxton's calendar the year 1484 began on Lady Day 25 March, the day before he finished printing *Aesop*; so when he says he wrote the translation 'in 1483' he may well mean the first two months of 1484 by modern reckoning. Probably he began it soon after completing *Cato IV* on 23 December 1483, for his work on *Knight of the Tower* and *Golden Legend* would hardly have left him time earlier in the year. If so, he must have set his cutter to work on the multitudinous illustrations two or three months earlier still, perhaps as soon as the *Canterbury Tales II* cuts were ready.

Caxton added two stories of his own to Poggio's, one rather rude, one edifying. The rather rude one is about a widow who decided to marry a certain widower, and was warned off by her officious maid. 'Alas,' said the maid, 'I am sorry for you, by cause I have heard say that he is a perilous man, for he lay so oft and knew so much his other wife that she died thereof, and I am sorry thereof that if ye should fall into like case.' To whom the widow answered and said: 'Forsooth I would be dead, for there is but sorrow and care in this world.' 'This,' remarks Caxton appreciatively, 'was a courteous excuse of a widow.' The edifying tale, 'which a worshipful priest and a parson told me of late', is of two priests who took their M.A.s together at Oxford. One 'that was pert and quick was anon promoted to a benefice or twain and after to prebends and for to be dean of a great prince's chapel'. Some years after, 'riding into a good parish with a ten or twelve horses like a prelate', he found his simple friend was parson there, and enquired how much it was worth. 'Forsooth,' said he, 'if I do my true diligence in the cure of my parishioners in preaching and teaching I shall have heaven therefore.' 'This was a good answer of a good priest and an honest,' says Caxton.

Meanwhile Richard's first and only true Parliament[2] had sat from 23 January to 20 February 1484. Among its acts was a restrictionist statute against Italian and other foreign merchants and craftsmen, with a curious proviso apparently inserted by the King himself or his Council, stipulating that the act should not apply to the importation or sale of 'books written or printed', or to the residence of alien merchants engaged in the book trade or 'any scrivener, limner, binder or printer of such books'. At first sight this enactment in the interest of foreigners might seem to injure the native Caxton (not to mention his friend the St. Albans printer) by favouring his alien rivals, William de Machlinia in London and Theodoric Rood at Oxford. But Caxton was probably no less dependent than they on alien craftsmen from

---

[1] Such a manuscript source seems much more likely than the lost printed edition postulated by R. H. Wilson.

[2] The scratch quasi-Parliament of June 1483, although legal enough, had been summoned for the crowning of little Edward V, and therefore ceased to exist when it proclaimed Richard king instead.

Germany or Flanders, and in any case his relations with contemporary presses seems to have been fraternal rather than competitive.[1] Richard had no previous cause to encourage this Woodville loyalist; however, in that same month of Parliament Richard was negotiating an entente with the Queen which, as we shall see, also narrowed the gap between himself and Caxton. The King may have been approached on behalf of Rood and the Oxford booksellers by Bishop Waynflete during his entertainment at Magdalen College on 24-25 July 1483.[2] William de Machlinia could claim support as a public benefactor for his series of *Year Books* and other essential law texts, including the *Statutes* of this very Parliament which he printed soon after. Perhaps all parties joined, together with the chief merchants of the book trade,[3] as a pressure group to secure this liberal proviso for their mutual advantage.

[1] The London and St. Albans printers used types modelled on Caxton's 2, 3, and 4, and presumably supplied by Veldener with Caxton's help. Rood in his *Festial*, 1486, used woodcuts apparently supplied by Caxton. All these specialised in Latin scholastic theology or grammar school texts, Lettou (who disappeared from the partnership early in 1483) and de Machlinia also in Law French, and only occasionally encroached on Caxton's chosen field of books in English. In the wideopen market of the 1480s there was room for all, and (as Veldener himself had found) it was more profitable to cooperate than to compete. Wynkyn de Worde and Pynson did likewise even a decade later.

[2] Waynflete, the founder of Magdalen, was addressed as patron in Carmelianus's commendatory verses for Rood's edition of the Magdalen School master Anwykyll's *Compendium grammaticae.*

[3] These then included the partners Henry Frankenberg and Bernard van Stondo, for whom de Machlinia worked, Peter Actors and Johannes de Westfalia, who supplied books to Hunte in 1483, and Thomas Hunte himself, the Oxford bookseller with whom Rood produced his 1485 *Phalaris.*

14

ORDERS OF CHIVALRY

O N 1 March 1484 Richard made public his pact with the Queen, negotiated during the Parliament and foreshadowed by Caxton's *Knight of the Tower*, by appearing in person before the Lords spiritual and temporal and the Mayor and Aldermen of London to swear *'verbo regio'*, on the word of a King: that if 'the daughters of Elizabeth Grey, late calling herself Queen of England' would only leave Sanctuary at Westminster, he would treat them 'as kinswomen', marry them to 'gentlemen born', and pay her warder Nesfeld 700 marks yearly for their mother's keep. Richard's motive was to counter Henry Tudor's Christmas vow to marry Princess Elizabeth, and also to take hostages for the Queen's future conduct. The terms are extraordinarily harsh, and incorporate acknowledgement of the nullity of the Queen's marriage and rank and the bastardy of her children. She, wretched woman, can only have been induced to surrender her last asset in the princesses to the murderer of her sons and brother, by despair at Henry's failure in Buckingham's rebellion, and by fear of worse. Richard's menaces can be identified by what he promises *not* to do: he will 'see that they shall be in surety of their lives' (i.e. not actually kill them), will not subject them to 'ravishment or defilement contrary to their wills', or imprisonment 'within the Tower of London or any other prison', or punish them on charges of treason 'before that they may be at their lawful defence and answer.[1]

In view of this stern bargain it is less surprising that Caxton presented his next book, the small undated quarto *Order of Chivalry*, 'to my redoubted, natural and most dread sovereign lord King Richard', with the same prayers for 'long life and prosperous welfare', with 'victory over his enemies' and 'everlasting life in heaven', as he had sent up in past years for Edward. Caxton translated from a French version of the Catalan *Libre del orde de cavayleria* by the Majorcan knight, mystic and missionary Raimon Lull (1235-1315). This devotional and unpractical treatise on the duties and symbolism of knighthood was given, according to Lull's fictional

---

[1] The terms imply that the Queen remained in Sanctuary, and there is no evidence that she came out before Bosworth. She would be a fool if she did. Richard could have her out if he chose for the major crimes of witchcraft (of which he had accused her in June 1483) or high treason (for conspiring with Margaret Tudor and the exiled Woodvilles). If her rank or sex spared her from burning or beheading she could be put away for life, like Eleanor Cobham or Margaret of Anjou. Parliament had just dutifully declared that Edward had married her 'by sorcery and witchcraft committed by the said Elizabeth', and that 'all their issue were bastards'.

setting, by an aged hermit to a young squire on his way to court to seek knighthood, who there presented it to the king. Caxton's prologue repeats this situation in real life: he translates the book 'at the request of a gentle and noble esquire, according to the copy that the said squire delivered to me', and he presents it to King Richard. Still more curiously, Anthony Rivers had appeared at his last fancy-dress tournament on 21 January 1478 'horsed and armed in the habit of a white hermit', and was himself guardian of a young squire, the Prince of Wales. Caxton must surely have translated *Order of Chivalry* at Rivers's command for presentation to Edward IV in honour of his ward and nephew the Prince. Now, when Edward was dead, and Richard had beheaded Rivers and murdered the little Prince, Caxton on behalf of Edward's widow, Rivers's sister, presented the book to the new King in honour of the new Prince of Wales.[1]

Richard, as a practical man of war, was a notorious lifelong non-participant in the makebelieve chivalry of Edward's reign. Yet Caxton makes bold to lecture him on the need for its revival. 'I would it pleased our sovereign Lord that twice or thrice in a year he would do cry jousts of peace [bloodless tournaments] to the end that every knight should have horse and harness and also the use and craft of a knight, and also to tourney one against one or two against two and the best to have a prize, a diamond or jewel, such as should please the Prince [i.e. Richard himself]. 'Caxton recommends the King to 'command this book to be had and read unto other young lords, knights and gentlemen within this realm, that the noble order of chivalry be hereafter better used and honoured than it hath been in late days past'. *If* Richard does so, 'herein he shall do a noble and virtuous deed, and I shall pray Almighty God for his long life' and so forth.

Caxton's censure of degenerate knighthood in the new reign echoes his outcry against the decadent City in *Cato IV* a few months earlier, and is even more daring in a commoner. 'O ye knights of England, where is the custom and usage of noble chivalry that was used in those days? What do ye now but go to the bains [baths, or rather brothels] and play at dice? . . . Alas, what do ye but sleep and take ease and are all disordered from chivalry? I would demand a question if I should not displease. How many knights be there now in England that have the use and exercise of a knight, that is to wit that he knoweth his horse and his horse him, his armours and harness meet and sitting, and so forth et cetera? I suppose and a due search

---

[1] Edward, son of Richard and Warwick's daughter Anne Neville, born in 1473, was created Prince of Wales on 24 August 1483, confirmed as heir to the throne by the 1484 Parliament, and died on or about 9 April 1484, thus extinguishing Richard's line as Richard had extinguished Edward's. It seems highly probable that Queen Elizabeth Woodville, when she surrendered her daughters to Richard on 1 March 1484, believed or was given to understand that Richard would marry Princess Elizabeth to the Prince as a last resort to forestall Henry Tudor. A year later Richard tried to marry her himself.

should be made there should many be founden that lack, the more pity is.' But he does not forget the opportunity to advertise again, as he had in *Godfrey of Boloyne* three years before, his future *Morte d'Arthur*, and perhaps an intended but never accomplished *Froissart*. 'Leave this,' he urges the delinquent knights, 'leave it and read the noble volumes of Saint Graal, of Lancelot, of Galahad, of Tristram, of Perceforest,[1] of Percival, of Gawain and many more. There shall ye see manhood, courtesy and gentleness. And look in latter days of the noble acts since the Conquest, as in King Richard's days Coeur de Lion, Edward the First and Third and his noble sons, Sir Robert Knolles, Sir John Hawkwood, and Sir Walter Manny. Read Froissart!' So *Order of Chivalry* was ingeniously designed by Caxton and its backers to be read as two-way propaganda, either for Richard or against him.

Order of Chivalry was probably printed soon after *Aesop*, towards April 1484, both in view of the political situation, and because it contains the large printed initial A which is found also in *Aesop* and was then perhaps lost or damaged, for it never appears again. Caxton's next dated book was *Morte d'Arthur*, 31 July 1485, his third longest work after *Golden Legend* and *Polycronicon*, which no doubt fully occupied the first half of 1485. The remainder of this gap is the only possible and suitable period for the production of the undated *Golden Legend*, which must therefore have been at press from about May 1484 to the end of the year.[2]

Golden Legend, a vast church and lay lectionary on the lives of saints, is Caxton's longest work and his only one on largest paper, which enabled him to print about 600,000 words in 449 leaves with double 55-line columns. The book is lavishly illustrated with 19 large page-width cuts of scriptural and hagiographic scenes, and 51 small column-width cuts of individual saints bearing their emblems, evidently by the *Game of Chess II* cutter, but supplemented by another hand, possibly the same as the *Aesop* interloper a little improved by practice. The enormous labour of translation took about fifteen months, from its first mention in *Polycronicon* in July 1482 to its completion on 20 November 1483. 'Forasmuch as this said work was great and overchargeable to me to accomplish,' says Caxton,[3] 'I feared me in the beginning of the translation to have continued it, by cause of the long time of the translation and also in the emprinting

---

[1] *Perceforest*, a fourteenth-century French prose romance popular in Burgundian Flanders, linked the Alexander the Great and Arthurian cycles. Perceforest is made King of Britain by Alexander, and his grandson brings the Holy Grail to England.

[2] The suggestion that the date 20 November 1483 in *Golden Legend* is that of printing must be rejected, both because the context shows that it refers only to completion of translation, and because Caxton's exceptionally large production in 1483 leaves no room for this huge book.

[3] Crotch inadvertently reverses the order of Caxton's two prologues, failing to notice that in the British Library copy, from which he transcribed, the first prologue is misbound after the second.

of the same, and in a manner half desperate to have accomplished it was in purpose to have left it after that I had begun to translate it and to have laid it apart, had it not been at the instance of the puissant, noble and virtuous earl my Lord William Earl of Arundel, which desired me to proceed and continue the said work and promised me to take a reasonable quantity of them when they were achieved and accomplished, and sent to me a worshipful gentleman a servant of his named John Stanney, which solicited me in my Lord's name that I should in no wise leave it but accomplish it, promising that my said Lord should during my life give and grant to me a yearly fee, that is to wit a buck in summer and a doe in winter, with which fee I hold me well content. Then at contemplation and reverence of my said Lord I have endeavoured me to make an end and finish this said translation, and also have emprinted it in the most best wise that I have could or might, and present this said book to his good and noble lordship as chief causer of the achieving of it, praying him to take it in gree of me William Caxton his poor servant, and that it like him to remember my fee.'

Old William FitzAlan, 9th Earl of Arundel (1417-87), was a Yorkist magnate of the old guard,[1] but had joined Hastings's cabal against the Woodvilles at the approach of Edward's death. Arundel's backing of *Golden Legend* in 1482 or early 1483 shows, like Hastings's patronage of *Mirror of the World* in 1481, that Caxton was prudently insuring himself with both sides before the coming struggle for power. In May 1483 Protector Gloucester renewed his appointment as Warden of the Cinque Ports,[2] so securing Arundel's adhesion after the liquidation of Hastings in June, and made him Master of Game south of Trent. It was in this capacity that Arundel made the honorific but rather unlucrative gift of venison to which Caxton playfully alludes; but no doubt Arundel's promise 'to take a reasonable quantity'[3] of copies formed the crux of the bargain. Above his narrative of the Earl's munificent intervention Caxton printed a huge halfpage woodcut of Arundel's badge and motto, a horse and oaktree lettered 'My Trust Is'. *Golden Legend* is Caxton's only book after *Aesop* in which he refrains from using his new printed initials and paragraph marks, and reverts to rubrication by hand, evidently because he or Arundel felt that his assignment to produce the book

---

[1] Arundel's career was significantly parallel to that of Caxton's other patron the still older Earl of Essex. Both fought in battle on the Yorkist side in 1461, were forced or persuaded to marry their heirs to Woodville sisters of the Queen, stood loyal to Edward against Warwick in 1469-71, helped to defeat the Kentish rebels in 1471, and served on the Council which ruled England for Edward during his absence in the 1475 invasion of France.

[2] This key post, to which Edward IV had appointed Arundel in May 1470 in place of the rebel Kingmaker himself, made him a natural patron for the Kentish Caxton.

[3] Similarly Richard had bestowed a buck apiece on the learned doctors John Taylor and William Grocin for their Latin speeches in his honour at Magdalen College Oxford on 25 July 1483.

'in the most best wise' required the splendours of manuscript coloration.[1] So, with *Order of Chivalry* and *Golden Legend*, Caxton charmed the support of the monarch and the topmost earl,[2] just as he had in the previous reign with Edward and Rivers; and in due course, with the Earl of Oxford under Henry VII, he would bring off the hat trick.

Golden Legend or *Legenda Aurea*—so called, as Caxton explains in the words of his French original, 'because like as gold passeth in value all other metals so this legend exceedeth all other books'—is an outsize compendium of saints' lives arranged in order of their feast days in the liturgical year, intended for priests to read aloud in divine service or sermons as well as for the general reader. Caxton used Jean de Vignay's French version as his main source, but translated five lives omitted by Vignay direct from the Latin original written by the Italian Dominican Jacobus de Voragine about 1260-7, and added twenty lives of English saints from the existing English translation of about 1430 called *Gilte Legende*. This text must surely have been unknown to him until after his own translation from Vignay's French was completed, since he would hardly have undertaken that colossal labour if this quite adequate English version had been available to him from the first. Caxton used a Vignay manuscript of the class in which narratives from the life of Christ have been segregated into a separate section. He is unique in following this with a companion sequence of Old Testament lives from Adam to Judith, using apocryphal and later as well as scriptural material, the latter half of which has been severely abridged by Caxton himself, or by his source, or by both. But this potted Old Testament is evidently a professional product; it is just the kind of addition that priestly scribes were constantly introducing into this ever expanding and evolving work, and calls for liturgiological authority and Latin Vulgate learning which Caxton did not possess. The prevailing hypothesis, that it is a do-it-yourself job by Caxton himself, seems unproven and unprovable, and it is much more likely that he found the text readymade in one of his manuscripts, except perhaps for the rather amateurish abridgement.

In *Golden Legend*, however, Caxton indulged himself more exuberantly than ever before, but for the last time, in his characteristic autobiographical insertions. He has seen the very foreskin cut from the infant Jesus in the Church of Our Lady at Antwerp, 'and there I know well that on Trinity Sunday they show it with great

---

[1] The rubricated initials in *Golden Legend*, and the capital spaces left to receive them, are larger and more numerous than in any other of Caxton's works. The British Library copy even has an opening initial illuminated in gold paint.

[2] Arundel served as cupbearer at Richard's coronation, and rode in his triumphal entry into London on 25 November 1483. In May 1484 (just when printing probably commenced) Richard made him commissioner of array to defend the southeast against the threat of invasion from Henry Tudor, and repeated the appointment in December 1484.

reverence, and is there borne about with a great and solemn procession and that though I be unworthy have seen several times'. Once, when they were riding together from Ghent to Brussels,[1] he was told by 'a noble knight named Sir John Capons', who 'had been Viceroy and Governor of Aragon and Catalonia and that time counsellor unto the Duke of Burgundy Charles', how King David composed his *Miserere* Psalm buried to the neck, 'till he felt the worms creeping in his flesh'. At Cologne he learned from 'a noble doctor' how an image of the Virgin spoke to St. Jerome in church, and was also informed that 15,000 men were martyred there at the same time as St. Ursula's 11,000 maidens. In the life of St. Augustine he adds the miracle of the child by the seashore, 'which I have seen painted on an altar of Saint Austin at the Black Friars in Antwerp'.[2]

Morte d'Arthur, which Caxton had announced four years before in *Godfrey of Boloyne* and again in *Order of Chivalry*, was the last of his big commissions from the era of Edward. Printing was finished on 31 July 1485, when Richard was at Nottingham waiting for Henry Tudor, only a week before Henry's landing at Milford Haven, and three weeks before Bosworth. The Queen's unnatural entente with Richard had reached its peak in the spring, when it seemed possible that Richard himself, after the convenient death of his wife Anne on 16 March, might marry Princess Elizabeth to produce an heir.[3] She took the idea so seriously that she even persuaded her son Dorset to leave Henry and risk his head with Richard. But Richard, dissuaded by his advisers, appeared in person before the Mayor and Aldermen and Court of Mercers to swear with virtuous indignation (his second public oath within a year) that he had not poisoned his wife, 'nor ever had the thought to marry in such manner wise'. Henry recaptured Dorset, and when he sailed for England left the shifty Marquess behind in pawn for the French King's loans. So now the Queen's hopes were again for the Lancastrian Henry Tudor; provided she could marry her daughter to a King of England, she no longer cared which. Once again Caxton reflects the political situation. For the first time since *Festial* two years before he avoids Richard's name by omitting the regnal year from the date of *Morte d'Arthur*;

[1] Possibly Caxton made this journey in the service of Duchess Margaret, who had a palace in each of these cities and usually resided in one or the other in 1468-76.

[2] When the Saint was writing *De Trinitate* he rebuked a child whom he found trying to ladle the sea into a hole in the sand. The child retorted that Augustine's efforts to understand the mystery of the Trinity were more impossible still, and so vanished.

[3] The situation extraordinarily resembled the death of Anne's sister, wife of Richard's brother Clarence, in 1476, at the moment when Clarence found it convenient to woo Mary of Burgundy. The nation had already taken it as a declaration of intent when at Christmas 1484 Richard produced Elizabeth at court in gowns that matched his wife's, and his wife promptly fell ill. Henry himself was so alarmed that he began negotiations for a second-string Welsh bride (a daughter of his boyhood guardian William Herbert Earl of Pembroke) in case Richard should marry Elizabeth.

for the first time he dares to mention Richard's unmentionable predecessor, by listing the Nine Worthies again, ending with Arthur, Charlemagne, and 'Godfrey of Boloyne, of whose acts and life I made a book unto the excellent prince and king of noble memory King Edward the Fourth'; and he describes how, soon after the publication of *Godfrey* in December 1481, he was 'instantly required to emprint the history of King Arthur' by a group of 'many noble and diverse gentlemen of the realm of England', who in that time and context could only be understood to belong to the Woodville party. Of these 'one in special' answered Caxton's objection 'that diverse men hold opinion that there was no such Arthur', by instancing Arthur's grave in Glastonbury Abbey, the impression of his seal 'in red wax enclosed in beryl' in St. Edward's Chapel at Westminster Abbey, Gawain's skull at Dover Castle, the Round Table at Winchester, and 'in Wales in the town of Camelot the great stones and marvellous works of iron lying under the ground, and royal vaults which diverse now living hath seen'.[1] Possibly, as Blake suggests, this 'one in special' was the chivalric Rivers himself, and it was he who supplied 'the copy to me delivered' which Caxton printed. Perhaps Caxton did not foresee that even Henry might have welcomed his book. The Tudor claimed collateral descent from Arthur King of Britain (not to mention Brutus and Cadwallader) on his Welsh father's side; he had vowed to name his first son Arthur after the once and future King; and next year he was to take his wife to Winchester so that the child should be born at the English Camelot.

Caxton's edition remained the sole source for Malory's text for four and a half centuries, until the sensational discovery in 1934 of a lost (or rather previously unidentified) manuscript in the Warden's bedroom safe at Winchester College. The true importance of the Winchester manuscript lay in its revelation of Malory's uncontaminated prose, as it was before the wellmeaning Caxton modernised his archaic diction and syntax, abridged or sometimes expanded his narrative, and reconstructed his eight original sections into 21 books and 507 chapters. The rest should have come as no surprise. Malory's French originals were extant and well known to modern scholars, especially to Professor Vinaver. Caxton himself announces that he has 'divided into xxi books and every book chaptered' a text which 'Sir Thomas Malory did take out of certain books of French and reduced it into English'. He speaks of the 'many noble volumes in French which I have seen and read beyond the sea, which be not had in our maternal tongue, but in Welsh by many and also in French and some in English, but nowhere nigh all', and explains that he is

---

[1] A reminiscence of Higden's account of the Roman remains then visible at Caerleon-on-Usk, borrowed from Giraldus Cambrensis, which Caxton had twice printed in *Description of Britain* and in *Polycronicon* bk. 1, ch. 48: 'And within the walls and without is great building under earth, water conduits and ways under earth, and stews [baths] also thou shalt see wonderly made with strait side ways of breathing that wonderly cast up heat . . . There the messengers of Rome came to the great Arthur's court if it is lawful to trow.'

printing only a mere sample of these, 'such as have late been drawn out briefly into English'. He called it *Morte d'Arthur*, with explicit caution and apology, from the title of the final book as given in Malory's colophon; it was the best title he had, and a much more appropriate one than *The Works of Sir Thomas Malory*. Professor Vinaver over-reacted, as though he had detected two fifteenth-century Englishmen in a conspiracy to mislead a twentieth-century French textual editor, Malory by writing 'a series of separate romances', and Caxton by producing it as a single book under a 'spurious and totally unrepresentative title'.

Owing to an accidental shortage in the first printing, two sheets (N3/6 and Y3/6) had to be reprinted, as usual with correction of a few obvious errors and commission of several new ones. Professor Vinaver deserves credit for discovering these variants, but not for misinterpreting this common mishap as an intentional revision, and misdescribing the second compositor's unmeaningful spelling preferences as 'nearly seven hundred corrections'.

During the fifteen months from May 1484 to July 1485, when *Golden Legend* and *Morte d'Arthur* occupied the whole resources of his press, Caxton had continued to write translations to await the completion of these major works. *Royal Book* was finished on 13 September 1484 but not printed until two years later. *Charles the Great* was completed in translation on 18 June 1485, *Paris and Vienne* on 31 August 1485, and both came off the press in December, after a gap in production of about three months due perhaps to the upheavals of Henry Tudor's victory.

Charles the Great, 1 December 1485, completed Caxton's promise made in Edward's reign, probably under commission from his Woodville patrons, to print the lives of the Three Christian Worthies Arthur, Charlemagne, and Godfrey. The French original, though anonymous, was a recent compilation by Jean Bagnyon of Lausanne and Geneva, consisting of his prose adaptation of *Fierabras*, a verse romance of the Charlemagne cycle telling how the good Saracen giant Fierabras was defeated and baptised by the paladin Oliver and eventually became a saint in heaven, sandwiched between a hardly more historical account of French kings up to the death of Charlemagne taken mainly from the *Mireur Historial* (a French version of Vincent of Beauvais's *Speculum Historiale*).[1]

Charles the Great was commissioned by 'some persons of noble

---

[1] The known dates in Bagnyon's career extend from 1463 to 1494. His authorship was first revealed in the edition printed by Jacques Maillet, Lyons, 21 July 1489. Caxton has been wrongly supposed to have used the edition of *Fierabras* printed by Louis Cruse, alias Garbin, Geneva, 13 March 1483, apparently through misunderstanding of the EETS editor S. J. H. Herrtage (1881), who merely states that *he* (Herrtage) has chosen to use that edition for comparison with Caxton's translation. In fact three other editions had appeared before Caxton's (Adam Steinschaber, Geneva, 28 November 1478; Simon Dujardin, Geneva, 1481; Guillaume Le Roy, Lyons, 16 November 1484). Caxton may have used any of the four or, equally possibly, a manuscript.

estate and degree', 'my good singular lords and special masters and friends'—presumably the same as the 'many noble and diverse gentlemen' who had requested *Morte d'Arthur*—and in particular by 'a good and singular friend of mine Master William Daubeney, one of the Treasurers of the Jewels of the noble and most Christian king, our natural and sovereign lord late of noble memory, King Edward the Fourth'. Caxton had ventured to recall Edward in *Morte d'Arthur* when Richard was still king, as a propaganda hint in support of Henry. Now in Henry's reign he did so again, perhaps as a reminder that Henry was expected to marry Edward's Woodville daughter, but perhaps also for a reason that was likely to please his patron Daubeney better than the new King. During Edward's three last years Daubeney had held not only the honourable and responsible post of Clerk or Treasurer of the Jewel House,[1] but also by a grant of 9 November 1480 the lucrative semi-sinecure of Searcher in the port of London.[2] Both posts were confirmed by Richard, who did not take on Edward's men for nothing;[3] he employed Daubeney to pawn the State jewels for City loans in December 1483, and on 24 April 1484 appointed him commissary general at the Admiralty under the gentle Brackenbury, with duties including defence against Henry Tudor's expected invasion. Immediately after Bosworth Daubeney was in trouble. He lost both his posts as a creature of Richard, and was implicated as 'then keeper of jewels of the said pretensed king' when the City dunned Henry to redeem the royal jewels which Daubeney had pawned for £20,000. He remained a disaffected Yorkist, and was beheaded on 3 February 1495 in a distinguished batch of Perkin Warbeck's supporters, including the King's step-uncle Sir William Stanley and Sir Simon Mountford. Perhaps Daubeney's fall in 1485, and Caxton's wellmeaning effort in *Charles the Great* to put in a good word for his 'singular friend', earned the Printer a black mark with Henry, and helped to cause his failure to find patronage in the new court during the next three years.

Charles the Great was immediately followed by *Paris and Vienne*,

---

[1] Modern scholars have confused him with his apparently unrelated namesake Sir William Daubeney (1424-60), a stay-at-home Somerset landowner who was the father of Henry's prominent warrior statesman Baron Giles Daubeney. In documents of 1480-95 Daubeney is called 'Master', 'gentleman, of Southwark', or at most 'Esquire', except that in the punitive Privy Seal of 1485 he is called Knight either in error or by his own unjustified claim. The *Black Book of the King's Household* under Edward provides that the Jewel Clerk's 'livery is as Knight's, and if he be sick he taketh in eating days [daily rations] like the Squires of the Body', but the post apparently did not carry actual knighthood.

[2] The Searcher was a customs official who inspected incoming and outgoing shipping for dutiable or contraband goods, receiving a percentage fee on their value and half of all confiscated wares, and performing his duties through deputies.

[3] On 16 December 1483 Richard awarded Daubeney arrears from 9 April, since when he had held his post as Searcher only 'by letter of Edward V the late bastard king'.

19 December 1485. This happily ending tale of courtly love between the young troubadour knight Paris and the princess Vienne belongs to the semi-realistic genre evolved in Provence and typified by *Aucassin and Nicolette*, in which the chivalric romance is about to become a historical novel; the action begins in a definite year, 1272, and shows intimate knowledge of the real town of Vienne in Dauphiny. Caxton translated from a later abridged version of a French translation made in 1432 by Pierre de la Cypède of Marseilles from a Provençal original. The French text was first printed not long after, by Gerard Leeu at Antwerp, 15 May 1487, from a manuscript close to Caxton's but not the same. Caxton and Leeu evidently had a special relationship, rather like Caxton's with Mansion, for the exchange of ideas and professional aid. Leeu, when he printed the Dutch *Reynard the Fox* in 1479 two years before Caxton's English translation of 1481, perhaps supplied Caxton with a copy of his own edition or of the manuscript from which he produced it. In 1492-3, soon after Caxton's death and just before his own, Leeu reprinted three Caxton editions in the English language (*Jason, Paris and Vienne*, and *Chronicles of England*), probably not by way of piracy, but rather as commissions from Caxton's executors or his successor Wynkyn de Worde to fill the gap until de Worde brought the press back into full production. It seems not unlikely that Caxton obtained his first set of printed initials early in 1484 from the same Gouda craftsmen who supplied the rather similar sets which Leeu took with him when he removed his press to Antwerp in the summer of that year, and that Leeu helped him to do so.

Paris and Vienne completed the series of four chivalric books which (along with *Golden Legend*) had been Caxton's mainstay during Richard's last fifteen months, and had probably originated in orders received from Woodville sources before Edward's death. Paradoxically his output had remained as prolific as ever in the uneasy climate of the usurper's reign. But he was living on his fat, and at the end of 1485, under a new dynasty, he had run out of commissions and patrons.

15

WITH A STRANGE DEVICE

HENRY'S victory of Bosworth on 22 August 1485 might have been expected to improve Caxton's fortunes. The traitor Richard perished, crying 'Treason, Treason' when the recreant Stanleys joined his enemies. Henry VII kept his vow by marrying Elizabeth on 18 January 1486; so the former Queen Mother was now a queen mother-in-law, the few surviving male Woodvilles seemed back in favour, and Caxton might hope to join them all in their second royal family. But in reality the new reign, though disguised as a happy union of the Red Rose and White, was a Lancastrian restoration, and marked the irreparable defeat of Yorkism. Henry never forgave or trusted his mother-in-law and her son Dorset after their desertion of his cause in 1484. Caxton himself did not turn his coat so easily, and it would be mistaken to see him as a Vicar of Bray, content to be servile to whatsoever king might reign. He had remained a lifelong Yorkist by ties and convictions, ever since the first beginnings of Yorkism in Duke Humphrey and Duke Richard forty years before; under Richard's usurpation he stood by Edward's captive widow, at his peril; and it was not until three years after, in 1488, that he first found or accepted Tudor patronage.

So it is not surprising that marked changes appear in Caxton's press after the end of 1485. The next two years, in which he printed only four or five substantial books and about as many smaller pieces, are his least productive period. His choice of texts shows that he had been obliged to find new customers. In 1486-8, instead of courtly romances, patriotic histories, merry tales, the English poets and his own translations, he turned to standard church or school books and lay devotional works. *Book of Good Manners*, 11 May 1487, is the only dated book of this period, and also the only one which is a new translation and has a named patron. Even in the ensuing years of partial recovery from 1489 to his death in 1491, Caxton produced only a single book with a date of printing[1] and only three with dates of completion of translation.[2] This shortage of dates, accompanied by a corresponding decline in his characteristic prologues and epilogues, presents formidable though not insoluble problems of chronology for the rest of Caxton's career, and is itself a symptom of malaise caused by lack of patrons and of incentives to translate. Caxton had always preferred to date and preface the books in which he took a special pride, because they were of his own composition and embodied his

[1] *Faytes of Arms*, 14 July 1489.
[2] *Doctrinal of Sapience*, 7 May 1489; *Art to know well to Die*, 15 June 1490; *Eneydos*, 22 June 1490.

gratifying and profitable relationships with patrons. Henceforth such books were in the minority.

Closer analysis shows, however, that Caxton's crisis, though intensified and prolonged by the changed conditions of the new reign, had its real origins under Richard. Already when Caxton began to print *Morte d'Arthur* early in 1485 he must have foreseen that his list would be all but exhausted by the end of the year. After *Charles the Great* and *Paris and Vienne*, the last of his hoard of chivalric orders held over from Edward's reign, he had nothing left in hand except *Royal Book*, translated in September 1484 for his unnamed mercer friend, which was the only new commission he had found under Richard. His two temporary patrons Arundel and Daubeney, both Edwardian Yorkists who had transferred their loyalty to Richard in order to keep the posts they had held under Edward, had only consented to subsidise one book each from his existing Edwardian list. The Queen Dowager, after her capitulation to Richard in 1484, no longer had any motive to patronise him. Caxton must have taken immediate action to secure his new clientele, and to order the requisite new types and woodcuts. So the changes of 1486 probably reflect Caxton's response to his situation several months before Richard's end, and have nothing to do with Bosworth.

Caxton's introduction of new types was equally unconnected with the advent of Henry. Type 4* was due to end its three years of useful life towards the close of 1485, and Caxton would have been obliged to order its replacement in any case no later than the midsummer of that year. By coincidence the moment was a convenient one in two ways. First, Caxton was able to use the opportunity to acquire type of new design to suit his new programme of church Latin texts, for which a dignified gothic was required, whereas a mere repeat of the vernacular 4* would have been inappropriate. Secondly Veldener, his supplier of type, was about to go out of business at Louvain, and Caxton was just in time to seize the last chance of using his services. By 1486 Veldener had printed his last books, and his types, including not only those he had used at his own press, but also recasts and modified versions, were being dispersed among a variety of printers at Louvain, Antwerp, and Delft.

The new type 5: 113G. was a reduced, lighter version of the gothic type 3: 135G., rather extreme in its heavy severity, which Caxton had used in Edward's time for liturgical texts. In this smaller and generalised form Caxton's new gothic was also quite presentable for vernacular printing, and in the lean years 1486-8 he used it for English books as well as Latin. As we shall see, it was not until 1488-9 that Caxton first introduced two companion types, the bâtarde type 6: 120B., which was a reduced version of type 2: 135B., and the small gothic type 7: 84G., which was a reduced version of type 5. Both these could hardly have been designed and cut by anyone but Veldener, and must therefore have been made before his cessation of business in 1486. Caxton, when ordering type 5 as a successor to type 3, would hardly have failed to order type 6 also as a successor to type 3's inseparable companion type 2. Probably all three types, 5, 6, and 7, were ordered at the same time towards mid-1485, and delivered by the end of the year. Then, when Caxton realised that his production would be restricted for some time to come, he economised by using only the dual-purpose type 5, and put the exclusively vernacular type 6 aside to await better days. Indeed, his diminished output in 1486-8 was evidently not sufficient to occupy all his presses; so if, as seems likely, he had hitherto been working four presses, he probably now continued on only two, and laid off half his workmen. Type 7 was suitable either for printing marginalia or as an Indulgence type. Perhaps he had ordered it for use in *Speculum vitae Christi*, his first book which requires marginalia; but in the event he used type 5 here instead,[1] and did not employ type 7 until his next order for Indulgences turned up in 1489.

The chronology of Caxton's 1486-8 books is problematical, owing to lack of dates; but it is not difficult to put them in a rational order which is probably near the truth. Perhaps *Directorium sacerdotum* came first, as it is a substantial Latin liturgical work of the kind for which type 5 was primarily intended, and new editions were already required in 1487 (printed by Leeu at Antwerp), and yet again in 1488 (Caxton's second edition). This is a revised version of the *Sarum Ordinal* which Caxton had produced for sale at the Red Pale, 'good cheap', about seven years before, when it was already long obsolete, for the *Directorium* was edited towards 1455 by Clement Maydestone, a Brigittine monk at Sion Nunnery near Isleworth in Middlesex, who died on 9 September 1456.

The only surviving copy of *Directorium sacerdotum I*, now in the British Library, contains a unique full-page folio woodcut *Image of Pity*, mounted on a front flyleaf at the time of the late eighteenth-century binding, but perhaps already present as a pastedown in the original binding. *Image of Pity* cuts, intended for sticking on walls, doors, boxlids, bookbindings etc., evolved from the favourite Missal illumination of the Mass of St. Gregory, which shows the Saint saying mass and seeing a vision of the dead Christ with the instruments of the Passion. Caxton's cut portrays Christ standing in the tomb, surrounded by a border of 28 compartments containing emblems such as the nails, ladder, scourges, pieces of silver, St. Peter's cock, the Pelican in her piety. The woodcut text has been mostly cut out, no doubt to prepare for a visit from Henry VIII's gestapo, but enough remains to show that it made the same inordinate promise of indulgence in the next world as a smaller version in Caxton's third edition of *Horae ad usum Sarum*: 'To them that before this Image of

---

[1] Caxton again printed the marginalia in type 5 in his second edition of *Speculum*, in 1490, although he had then already used type 7 in his 1489 *Indulgence*. Wynkyn de Worde reprinted *Speculum* in 1494, and then at last used Caxton's type 7, which he had inherited, for its probable original purpose in the marginalia.

Pity devoutly say five Paternosters, five Aves and a Credo, piteously beholding these arms of Christ's Passion, are granted 32,755 years of pardon.'[1]

This *Horae III*, an octavo, survives only in an eight-leaf fragment comprising quire m, and containing the above-mentioned smaller version of the *Image of Pity* cut. Four more woodcuts from the same *Horae* series occur in Caxton's *Speculum vitae Christi I*, and many others in *Horae* and other books printed with Caxton's materials by Wynkyn de Worde after Caxton's death. A perfect copy of *Horae III* would no doubt include the whole set of thirty-seven cuts or most of it, and if we can assume that the *Horae* cuts were first used in the book for which they were originally intended, then *Horae III* must have been printed before *Speculum I*.[2] Similarly *Speculum I* in turn is illustrated with its own series of woodcuts of the life of Christ, and a few of these are found also in *Royal Book*; so apparently *Royal Book* came after *Speculum I*.

Speculum vitae Christi, a life of Christ with devout meditations, was universally attributed to St. Bonaventure (1221-74),[3] but is now considered to be the work of another Franciscan, Johannes de Caulibus, of San Gemignano, the city of beautiful towers near Sienna. Nearly sixty editions, mostly in Latin or Italian, were printed before the end of the fifteenth century. Caxton reproduced the English translation made directly from the Latin[4] by Nicholas Love, Prior of the Carthusian monastery of Mount Grace of Ingleby in Yorkshire, who presented it in London to Thomas Arundel Archbishop of Canterbury in 1410. Arundel recommended it 'for the edification of the faithful and the confutation of heretics or Lollards'; and indeed the text includes many interpolations expressing Love's hate of the unfortunate Lollards, who were still occasionally burned under both Edward and Henry. Caxton illustrated the book with a sequence of twenty-five woodcuts of the life of Christ,[5] eked out with

[1] Two other versions of *Image of Pity* were probably produced by Caxton, one impressed on a blank leaf in a book printed by Mathias van der Goes at Antwerp c. 1490, *Colloquium peccatoris et Christi*, in Cambridge University Library, with the text woodcut in imitation of Caxton's type 5, and the other in Wynkyn de Worde's first edition of *Sarum Horae*, c. 1494 (Duff 182), in which the woodcuts are from Caxton's stock. Various other English and foreign versions are known. The accompanying purported Indulgence was unauthorised by the Church.

[2] It is possible of course (as Hodnett suggests) but less likely that Caxton used *Horae* cuts in *Speculum I* before printing *Horae III* itself. Whichever happened it still seems reasonable to date *Horae III* to 1486 or 1487, and not as Duff does to 1490.

[3] Blades says the book was written 'by Saint Bonaventura, in 1410'!

[4] And not, as Blades implies, from the French version made about 1422 by Jean de Gallopes (the same whose prose rendering of *Pilgrimage of the Soul* was wrongly supposed by Blades to have been used for the translation printed by Caxton).

[5] The only surviving copy of *Speculum I* in Cambridge University Library is imperfect, but a complete copy would no doubt contain all the twenty-five cuts from the original series which are found in Caxton's second edition, in 1490. Hodnett has identified three more cuts from the series in later books.

four more from the *Horae III* series. Probably he had ordered both sets from Flanders in 1485 at the same time as his new types, but from different craftsmen, for the *Speculum* cuts are northern French in their simple elegance, while the crude, quaint *Horae* cuts evidently come from the Dutch side of Flanders. Apparently he had dismissed his own woodcutters, for no new work in their cheerfully incompetent style is found after the 1484 *Golden Legend*.

The undated *Royal Book*, in which Caxton re-used six more or less appropriate *Speculum* cuts,[1] was presumably printed early in 1487, immediately after *Speculum I* (1486), and just before *Book of Good Manners*, 11 May 1487. Caxton's translation, which he had completed long before on 13 September 1484, came from *Livre des vices et des vertus*, also called *Somme le Roi*, written in 1279 for Philippe III le Hardi King of France[2] by his Dominican confessor, Frère Laurent, as a lay handbook on 'the Ten Commandments of Our Lord, the Seven Deadly Sins with their branches, the Seven Virtues and many other holy things and matters good and profitable for the weal of man's soul'. *Book of Good Manners* is Caxton's translation, completed on 8 June 1486, of a similar treatise on the vices and virtues as practised by the three social orders of Church, nobility and common people, the *Livre des bonnes moeurs* by the Paris Augustinian Jacques Legrand, who presented it to Jean Duc de Berry on 4 March 1409.[3] 'An honest man and a special friend of mine,' says Caxton, 'a mercer of London named William Pratt which late departed out of this life, on whose soul God have mercy, not long before his death delivered to me a little book in French called Book of Good Manners . . . and desired me instantly [urgently] to translate it into English our maternal tongue to the end that it might be had and used among the people for the amendment of their manners', which Caxton did 'at the request and desire of him which was my singular friend and of old knowledge'. It has often been suggested that Pratt was also the unnamed 'singular friend of mine, a mercer of London' for whom Caxton had translated *Royal Book*.[4] A closer look at the circumstances seems to confirm this view. The mercer William Pratt, with whom Caxton had taken livery and played truant from the Lord Mayor's procession long ago in 1452, had died in July 1486, a few weeks after Caxton finished translating *Book of Good Manners*. His

[1] Blades mistakenly says the cuts are here used 'without any reference to fitness', but in fact a cut of the Blessed Sacrament is selected to illustrate that subject, and so on.

[2] Not, as Caxton (or his manuscript source?) says, for his successor Philippe IV le Bel.

[3] Duff mistakenly asserts that Caxton translated the edition printed by Pierre Le Rouge at Chablis, 1 April 1478, which in fact has a different text. No doubt Caxton used a manuscript, as usual.

[4] Pratt was also possibly the 'singular friend and gossip of mine' for whom Caxton printed *Boethius* in 1478, as Crotch suggests. Crotch's remark that 'it was a group of friends, this time "mercers of London", not nobles, who inspired Caxton to print the *Royal Book* in 1488' can only be due to a slip of memory, for Caxton says nothing of the sort.

will—a pious one requesting a funeral 'in the most lowliest and devout wise', with 'neither cloth of gold nor cloth of silk laid on my body to buryingward, but only so much black wool cloth as shall suffice to make two gowns, and a cross of white linen'—was proved on 4 August 1486. The gap of two and a half years between translation and printing of *Royal Book* is unique in Caxton's career, but was evidently caused by the huge task of producing *Golden Legend* and *Morte d'Arthur* in 1484–5 and by the priority of Caxton's stopgap commissions in 1486. He at last printed the two books together in the spring of 1487, partly because this was his first convenient opportunity, partly in deference to his friend's memory, but also perhaps because Pratt's executors would be obliged to pay the bill.

Type 4*, although laid aside since December 1485, was fortunately not too worn for emergency use. It was no doubt in the latter half of 1487 or the first half of 1488 (since he would have been too busy before or after) that Caxton reprinted the greater part of *Golden Legend* in a new setting, matching the first printing of 1484 as closely as possible by using the same woodcuts, omitting all initials and paragraph marks for the rubricator to supply, employing the same type 4* for the text, but substituting for type 3 in the headings the smaller and neater type 5, so that one can instantly recognise any leaves which belong to the second printing. This partial reprint does not deserve the status of a second edition, which Blades, Proctor, Duff and others have given it, for it includes little more than half of the work, and was evidently intended to make up deficiencies in the stock of unbound sheets for the first printing. Only a minority of surviving copies are 'pure' examples of the first printing; the majority include varying numbers of sheets from the second printing. The usual hypothesis of accidental miscalculation of the numbers required, though it accounts for Caxton's frequent recourse in other books to the reprinting of single sheets, is inadequate to explain such a gross shortage. Some serious accident of fire or flood must have destroyed or spoiled most of Caxton's stock of sheets of more than half of the first printing. It is remarkable that exactly four sevenths (256 of 448 leaves) of the work had to be reprinted, while the remaining three sevenths is found only in the first printing. Apparently the first printing was distributed between several presses in seven batches which were stored separately, and of which only three escaped damage.[1]

Two other smaller folio pieces in type 4* were perhaps printed during the same temporary reappearance of this type. *Deathbed Prayers* is a single leaf printed on one side only, containing prayers

---

[1] An alternative explanation would be an unforeseen shortage of the necessary large size of paper, which might have prevented Caxton from completing the required number of copies at the time of the 1484 impression. A similar mishap occurred in the production of Colard Mansion's large-paper *Ovid* in the same year, 1484, and necessitated the reprinting of exactly the same proportion, four sevenths (28 of 49 quires)!

beginning 'O Glorious Jesu' to be recited by the dying. *Life of Saint Winifred* tells how the national woman saint of Wales was beheaded on Sunday 23 June 636 by young Prince Cradock while defending her virtue. The wouldbe ravisher, as he 'stode and wiped his sword on the grass', fell suddenly dead, turned black, and was carried away by fiends; but Winifred lived for fifteen years more with her head miraculously restored, and 'only a little redness in manner of a thread' to mark the join, and her body was 'translated' in 1138 from Holywell in Flintshire (where her cult was revived by Baron Corvo and persists to this day) to Shrewsbury Abbey. Caxton used an abridged version of her Latin life by Robert of Shrewsbury (about 1140), 'reduced into English by me William Caxton',[1] and included the Latin text of her feastday services as celebrated at Shrewsbury Abbey. Probably the book was commissioned by Shrewsbury Abbey for sale to pilgrims, in expectation of increased custom and royal encouragement in view of the Welshness of Henry Tudor and his welcome at Shrewsbury on 15 August 1485 on the way to Bosworth. Caxton may have owed this order to the Benedictine Abbey connection which had already served him at Westminster, Abingdon, and Saint Albans;[2] and perhaps it earned him a first faint good mark with the new dynasty.

Caxton's folio edition in type 5 of the fourth-century Aelius Donatus's Latin grammar, in the recent revision by the Italian humanist Antonius Mancinellus known as *Donatus melior*, is known only from two fragmentary leaves on vellum discovered in 1893 by Robert Proctor at New College Oxford.[3] Perhaps the finding-place of this fragment is no coincidence, and Caxton's edition may have been ordered by the University and its booksellers; for the Oxford printer Theodoric Rood had gone out of business after producing an edition of Mirk's *Festial* on 14 October 1486. Indeed 1486, whatever the reason, was a fatal year for English printers, and saw the end not only of the Oxford press but of the Schoolmaster Printer at St. Albans and of William de Machlinia in London.[4] During the five years until his death in 1491 Caxton remained the only printer in England.

It was natural that Caxton should decide to crown his new liturgical

[1] So *Life of Saint Winifred* is one of the only two texts which Caxton himself claims to have translated from Latin, the other being the life of St. Roche in *Golden Legend*.

[2] And perhaps Winchester also, for *Morte d'Arthur*?

[3] Nearly 400 editions of Donatus were printed within the fifteenth century, often on vellum for durability. Most were worn to pieces by schoolboys or survive only in fragments found in book bindings.

[4] The St. Albans press vanished in 1486 after printing *Book of Hawking, Hunting, and Blasing of Arms*, in which a worn remnant of Caxton's type 3 is used for headings. De Machlinia's last datable work is Pope Innocent VIII's *Bull* ratifying the marriage of Henry VII and Elizabeth of York, promulgated on 27 March 1486 and no doubt printed very soon after. Duff's view that he continued printing until 1490 is untenable, being founded on Proctor's misdating of *Promise of Matrimony* as 'after March 1486' instead of early 1483.

connection by publishing the most indispensable and therefore profitable service-book of the English Church, which hitherto had been produced only in expensive and laborious manuscript. The *Sarum Missal* was in greater demand than any other single book in pre-Reformation England, for every mass-saying priest and every church or chapel in the land was obliged to own or share a copy. Instead of printing it himself, however, he commissioned a French printer, Guillaume Maynyal, who completed the edition 'at the expense of the excellent man William Caxton' in Paris on 4 December 1487. Why did Caxton go to the trouble and expense of having this book printed abroad? Evidently he was compelled by his technical inability to execute the high-quality redprinting which was necessary for the rubrics, and by memories of his unsuccessful attempt at one-pull redprinting in the days of Colard Mansion at Bruges. Once again Caxton was first in the field, for this is not only the first edition of the *Sarum Missal*,[1] but the first Paris-printed Missal to contain redprinted rubrics.[2] To emphasise his property in the book Caxton impressed his wellknown printer's device, which is here used for the first time, on the blank final page after the printed sheets reached Westminster.

Caxton's device again appeared, though without his name, in Maynyal's equally splendid edition of *Legenda ad usum Sarum*, Paris, 14 August 1488. Duff in 1905 brilliantly predicted the existence of this book from fragments in Cambridge University Library, but it was not until 1956 that the unique surviving copy was discovered by Paul Morgan when cataloguing the ancient library of St. Mary's Church, Warwick. The *Legenda* is an abridged version of the Breviary, containing the Lectionary or readings from the Scriptures, lives of Saints and Homilies for each day of the liturgical year, but omitting the Antiphoner which gives the anthems and psalms.

Caxton gave a *Sarum Missal* to Westminster Abbey, where it appears in the 1520 inventory of St. Edward the Confessor's chapel as 'a paper massbook of Salisbury use of William Caxton's gift', and bequeathed at least sixteen *Legenda* to his parish church of St. Margaret's Westminster, which sold them between 1496 and 1500 at prices from 6s 8d downwards. In a disputed bequest to his son-in-law Gerard Crop in 1496 'twenty printed Legends' were valued at 13s 4d each, and no doubt this is nearer the published price.

[1] An undated edition printed by Michael Wenssler at Basel was ascribed to 1486 by Duff, but no doubt belongs to 1489, when on 17 November a passport was issued to Wenssler and others for conveying 'one small and four large barrels of printed books' down the Rhine to England.

[2] Previous Paris Missals, beginning in 1481, used black-printed type of a different size for the rubrics. Maynyal must have learned the technique of liturgical redprinting elsewhere, either at the then dominant centres of Basel or Venice, or at Lyons where it was first practised in 1485. Maynyal had previously printed in 1480 in partnership with Ulrich Gering, who with his fellow Germans Crantz and Friburger had founded the first press in France at Paris in 1470. Duff and others mistakenly supposed that Gering's partner was a different Maynyal, that his name was George, and that both were English.

5 Caxton's Device, from Legenda ad Usum Sarum, 1488.

It is tempting to connect Caxton's importation of the *Missal* and *Legenda* in 1488 with Miss N. J. Kerling's discovery of entries in the London customs accounts which show Caxton importing 112 books worth £13 on 25 February 1488, and on 25 April an unspecified number worth £10 16s 8d and 1049 worth £17 5s 0d.[1] Some have thought that these entries imply a lifelong activity by Caxton in the cross-channel book trade; but the fact that Caxton's name has not been found in the customs accounts in any other year during the whole period 1476-91 strongly suggests that these were once-only transactions relating to his business with Maynyal.

Caxton must have ordered the woodblock for his device specially for his two Maynyal books, though he continued to use it for the rest of his career. I used to think one might detect in it the heavy hand of his *Game of Chess II* cutter, but I now see that the decoration of the borders, with the interlocking foliage, flowers, and white on black, is unmistakably in the style of the Gouda woodcutter who supplied initials not only for Gerard Leeu, Gotfridus de Os and other Gouda printers, but also for Caxton himself at this time and again two or three years later (in a series used only after his death by Wynkyn de Worde). So Caxton ordered the device and the new initials together from Gouda in 1487.

Caxton's device, like those of many other fifteenth-century printers, shows his personal merchant's mark with his initials,[2] and is closely paralleled by other marks of the period in England and Wales.[3] Any contemporary would certainly interpret the device as Caxton's mark as a mercer; but he would also recognise that the lines are here so drawn and orientated (though not so in other examples) as to read alternatively 47 or 74.[4] This must surely allude to two crucial years in Caxton's career, 1447, which would presumably refer to his obtaining the freedom of the Mercers' Company,

[1] The valuations are surprisingly low, varying from an average of 2s 4d to only 4d per volume, but are similar to many others found by Miss Kerling for shipments by other book importers. No doubt they reflect merely a technical peculiarity in the customs clerk's method of assessment, rather than the actual price per volume to the purchaser. Caxton also exported 140 volumes in French, valued at £6, on 10 December 1488, possibly (as Miss Kerling suggests) unsold remainders of his Bruges editions.

[2] Caxton no doubt bore in mind the devices of his master Veldener and his colleague Mansion. Veldener's showed his merchant's mark and surname, Mansion's bore his initial M with a crescent moon as a rebus of his surname, the Flemish spelling of which (*manchoen*) meant just that.

[3] Blades reproduces that of 'John Felde, Alderman of London, merchant of the Staple of Calais. Dyed m. cccc. lxxvij' from a brass in Standon church Herts., but does not justify his claim that Felde was a mercer. Two more, perhaps carver's marks, were found by the Rev. Maurice H. Ridgway, F.S.A., on late fifteenth- or early sixteenth-century screens in the churches at Llanengan, Caernarvonshire, and Llanegryn in Merionethshire.

[4] The arabic figure four was then written in this form (though the modern form was also current, and was adopted in Caxton's types 2*, 4, and 4*, which are his only types with arabic numerals). The punctuation ('.4.7') would give priority to 47, but without precluding 74.

and 1474, when he issued his first book as a printer.[1] It would be just like Caxton when designing his device in 1487 to adapt his existing mercer's mark into this numerological pun. So it seems that the two apparently incompatible explanations of Caxton's device—as a typical merchant's mark, and as a highly personal pair of dates—are both applicable and true.

Crotch proposed a third interpretation, that a certain William Caston of Calais was the same man as the Printer, and that Caxton's device was copied from his seal. There is no doubt about the real existence of this William Caston, who appears in two documents of 1452 and 1460 as a property-owner in Calais, and in 1471 as the possessor of a manuscript volume with Calais connections; but a closer look shows that his identification with the Printer is hardly tenable. The first document is a power of attorney from Alexander Childe, 'citizen and clothier of London' to deliver seisin of two cottages and adjacent land in Calais to five persons, including 'William Caston of Calais esquire', dated 17 April 1452. The second is a lease of the same property to Lodowic Lyneham alderman of Calais by four persons including the same 'William Caston of Calais', dated 7 November 1460. The book, now in Boston Mass. Public Library, is a tract volume in which most items are written by a scribe signing himself 'V.C. Moris, native of Calais', and several are of local Calais interest. William Caston's ownership signature occurs three times, once alone, and twice with the date 1471 and a note that he has given the book to William Sonnyng.[2] This resident and man of property in Calais, who writes in an awkward, plodding hand, consistently spells his name Caston, and has the gentleman's or soldier's title of esquire,[3] can hardly be the same as our hero, who then resided at Bruges, wrote more millions of words than any other Englishman of his time, insisted on spelling his name Caxton and never appears as Caston even in documents, and was no gentleman, but always styled 'merchant, of London'. Caston of Calais was no doubt yet another distant relative and namesake in the Caxton-Causton-Caston clan, of whom we have already found a dozen bearing the favourite family christian name William. The Printer, whose occasional minor contacts with Calais as merchant and diplomat we have seen, may well have known him personally.

What of the seal? The document of 7 November 1460 bears five seals, one of which shows an eagle's head erased (the heraldic erasure or severance from the body being indicated by the conven-

---

[1] *Recuyell* appeared in 1474 by Caxton's Flanders dating, but in the spring of 1475 by the modern calendar.

[2] Blake suggests that the entries may be in Sonnyng's hand, but they are certainly in Caston's, as the hand is the same in the entry in which he writes only his own name.

[3] So he was not only not a stapler, as Crotch and others call him without authority, but presumably not a merchant at all. His title of esquire would fit a customs official (like Caxton's patron and friend William Daubeney esquire in the port of London) or an officer in the garrison of Calais.

tional three jagged projections below the neck), between the initials WC. But Crotch's 'striking resemblance' to Caxton's device does not exist. The eagle's head is delicately and even naturalistically engraved, and in the crude outline diagram with which Crotch purports to bring out the fancied relationship it is difficult to see any likeness either to the seal or to the device. A more careful inspection of the original document reveals that the seal is not Caston's at all! The document, as is usual with medieval deeds of this kind, is sealed by the witnesses, not by the parties to the transaction. In fact it is attested by the Mayor and various aldermen of Calais, the seals are theirs, and the eagle's head and initials WC belong to the Calais alderman William Cresse.

But the most surprising and significant feature of Caxton's device was not intended by the Printer and has not previously been noticed. As we shall see, it can be used as a clue to the chronological labyrinth of his last years, as a precise instrument for dating the ten undated books in which he used it.

16

THE TWO QUEEN MOTHERS

NO known edition from Caxton's own press is dated 1488, and no positive reason has yet been produced for assigning any book to that year, except the *Legenda* printed for him by Maynyal. But if we examine the progressive damage in the bottom frameline of Caxton's device, we shall find the missing evidence. On its first appearance in *Sarum Missal*, 4 December 1487, this frameline already shows a 10 mm. gap at the righthand end. In *Sarum Legenda*, 14 August 1488, another 10 mm. gap occurs 6 mm. to the left of the first. During 1489 two further stages of damage occurred; so it seems reasonable to assign the only two books which contain the device in the same state as in *Sarum Legenda* to 1488. These are second editions of Maydestone's *Directorium sacerdotum* and *Reynard the Fox*, and must be the first books printed in Caxton's bâtarde type 6: 120 B., which was therefore in use a year before its first dated occurrence in *Faytes of Arms*, 14 July 1489. As we have seen, type 6 is a reduced and simplified version of the Burgundian type 2: 135B., and was probably supplied along with type 5 by Veldener in 1485–6, but laid aside by Caxton during the two first lean Tudor years. He must indeed have been eager to use it at last, as the vernacular type 6 is not really appropriate to the liturgical Latin of *Directorium sacerdotum*, for the first edition of which he had made a point of inaugurating his gothic type 5. Two other unexpected features of *Directorium II* appear in the page containing the table for finding Easter, in which Caxton not only used his old type 4* for the very last time, but printed in red and black for the first time at Westminster, not by the faulty single-pull method he had used at Bruges, but by the correct two-pull technique,[1] no doubt spurred on by the success of Maynyal's recent red printing in the *Missal* and *Legenda*. This table covers the period 1489–1593, which suggests that *Directorium II* was produced late in 1488 for sale in the following year, and therefore that *Reynard the Fox II* came first.

For *Reynard II* Caxton wrote a new epilogue, which breaks off in mid-sentence in the only surviving copy in the Pepysian Library at Magdalene College Cambridge, but is supplied in seventeenth-century manuscript. The story, Caxton remarks in jesting earnest, does not reveal how Reynard died, 'but I ween he was hanged, for he highly deserved it, for he was a shrewd and fell thief and deceived the

[1] Red first, then black. Blades says: 'the black form being first printed, and the red form secondly and separately', but in fact the black is visibly printed over the red, as was the invariable rule in fifteenth-century redprinting, and this is shown correctly even in Blade's own facsimile of the *Directorium* page.

King with lies, and so mote all false traitors and such as be plained with any villainy be hanged by their necks, I should be well apaid. Yet there be many such which nevertheless abide in great worship all their lives, yet that helpeth not but that they go to hell when they die, and the devils pull them by their beards and burn their arses with hot irons ... God grant us his grace that we may not come thereunto, it growleth me sore and mine hair standeth right up when I think thereon.' This otherwise lost epilogue is evidently authentic, and shows that Caxton had at last resolved to buy patronage in exchange for propaganda from the Tudors as he had from the Yorkists.[1] Caxton's 'false traitors' were not far to seek. Richard's official heir the Earl of Lincoln, after pretending reconciliation to Henry, had been defeated and slain at Stoke on 16 June 1487 in the cause of three simultaneous claimants to the throne—the impostor Lambert Simnel pretending to be Clarence's son Edward Earl of Warwick, the real Edward whom Henry held in the Tower, and Lincoln himself. Caxton's former patroness Duchess Margaret of Burgundy had acknowledged Simnel as her nephew, and financed their army. Another early patron of Caxton had been implicated in the abortive rising of Richard's Lovell 'our Dog' and the Stafford brothers at Easter 1486. The Staffords fled into sanctuary with Abbot Sant at Culham Priory, a subsidiary of Abingdon; but Henry, less longsuffering than King Noble the Lion in the tale of Reynard, dragged them out and hanged Sir Humphrey. Sant was summoned before King's Bench on 28 June to show evidence of sanctuary rights. Henry forgave the Abbot by appointing him to a diplomatic mission to Brittany on 17 March 1488; but the old fox Sant persisted in his Yorkist intrigues, was convicted of high treason by the Parliament of January 1490 for plotting the release of Clarence's son, pardoned on 2 February 1493 on condition that he said a daily mass for the King's welfare (let us hope he did not forget), and remained Abbot of Abingdon until his natural death in April 1496.[2] Here was at least one false traitor who abode in great worship.

The King himself, as we shall see, had already become Caxton's patron by January 1489, through the good offices of the Earl of Oxford; so it seems legitimate to assign to the latter half of 1488 the book, *Four Sons of Aymon*, in which Caxton tells how he gained the patronage of Oxford. Earlier still, at the request of Oxford, 'my

[1] The authenticity of the *Reynard II* epilogue has been questioned but not discussed. It is shown by the unmistakably Caxtonian style, humour, and political context, and clinched by the nonce word *growleth*, Caxton's typically Humpty-Dumptyish coinage to represent the Middle Dutch verb *growelen*, used impersonally, meaning 'it horrifies me', which occurs twice previously in his *Reynard* translation, but apparently nowhere else in the English language.

[2] Modern historians are therefore mistaken in saying that Sant was 'hanged in December 1489'. Capital punishment for a high church dignitary was unthinkable until the reign of Henry VIII. Henry VII merely condemned Sant to forfeit 'his possessions not belonging to the Abbey', and abolished right of sanctuary in cases of treason.

singular and especial lord' (as Caxton calls him, by a title he had given in the past to Rivers and Arundel),[1] he had 'reduced and translated out of French into our maternal tongue the life of one of his predecessors named Robert Earl of Oxford toforesaid with diverse and many great miracles, which God showed for him as well in his life as after death, as it is showed all along in his said book'. The last phrase, which Caxton habitually uses when recommending a work to his public, suggests that *Life of Robert Earl of Oxford* was actually printed, although no trace of the book has yet been discovered either in print or manuscript.[2] Next Oxford 'late sent to me a book in French . . . *Les quatre fils Aymon*, which book according to his request I have endeavoured me to accomplish and to reduce it into our English to my great cost and charges as in the translating as in the emprinting of the same, hoping and not doubting that his good grace shall reward me in such wise that I shall have cause to pray for his good and prosperous welfare'.[3] The romance of Renaud of Montauban, and his magic horse Bayard which carried all four brothers at once (rather like Tom Pearse's grey mare) into battle against Charlemagne, was appropriate to both printer and patron. Caxton, when he began to translate *Recuyell* nineteen years before, 'took pen and ink and began boldly to run forth as blind Bayard',[4] and gallant Oxford fought his battles on the same principle, losing Barnet for Warwick but gaining Bosworth and Stoke for Henry. Even so theirs was a strange conjunction, and displays Caxton's characteristic ability to charm great men, or to convince them of the power of the press. Caxton's hero Tiptoft had beheaded Oxford's father and brother in 1462, and Oxford beheaded Tiptoft in 1470, fought on the opposite side to Caxton's patron Earl Rivers at Barnet 'in the time of the troublous world', and raided England twice in 1473-4 in collusion with false Clarence, for whom Caxton under Duchess Margaret's wing was then about to write his first propaganda prologue in *Game of Chess I*.[5] After ten years' incarceration in

---

[1] Rivers in *Moral Proverbs* ('my special lord'), Arundel in *Golden Legend* ('my special good lord').

[2] Even the subject remains unidentified. Robert the third Earl (1164-1221), a participant in Magna Carta and a benefactor of the Benedictine Priory at Hatfield Broadoak in Essex where he was buried, seems more likely than the ninth Earl (1361-92), the rather disreputable favourite of Richard II, who would not be a grateful memory under a Lancastrian restoration. Two lesser Roberts, the fifth (1240-96) and sixth (1257-1331), were also benefactors of the Church.

[3] This semi-humorous plea for payment was quite in order in the etiquette of patron and protégé, and Caxton had reminded Arundel in the same way 'that it like him to remember my fee'.

[4] Curiously Bayard was not blind in the romance, and the origin of the proverb remains unknown to this day. Caxton could be quoting from any of his three favourite poets, Chaucer (*Ye been as bold as Bayard is the blind That blundreth forth, Canon Yeoman's Tale* 860), or Gower or Lydgate (in his *Troy Book* which Caxton had then just reread), who both echo Chaucer.

[5] Oxford must have been deeply involved with Clarence. The Clarence-inspired Cambridgeshire rising in May 1477, which led to Clarence's arrest, was led by an impostor pretending to be the imprisoned Oxford (just as Simnel

the fortress of Hammes near Calais he had escaped in August 1484, taking his own gaolor Sir James Blount with him to join Henry Tudor against Richard. Now Oxford was the grandest magnate in the land, like Rivers before him. By an irony of history one of his tasks was to keep Clarence's son Edward in the Tower, and also, when Henry thought the time had come, on 28 November 1499, to behead him. As it turned out, the prudent Henry chopped off far more heads than the villainous Richard, or even the amiable Edward.[1]

At the same time Caxton made or renewed the equally exalted conquest of 'my Lady Margaret, mother unto our natural and sovereign lord and most christian King Henry the Seventh', to whom he dedicated his edition of the thirteenth-century romance *Blanchardin and Eglantine*: 'which book I late received in French from her good grace, and her commandment withal for to reduce and translate it into our maternal and English tongue, which book I had long tofore sold to my said lady, and knew well that the story of it was honest and joyful to all virtuous young noble gentlemen and women for to read therein as for their pastime. For under correction in my judgment it is as requisite . . . for gentle young ladies and damsels for to learn to be steadfast and constant in their part to them that they once have promised and agreed to, such as have put their lives oft in jeopardy for to please them to stand in grace, as it is to occupy them and study overmuch in books of contemplation.' Caxton's political allusions, although undetected by modern scholars, were clear enough to his readers. His significant 'long tofore' can only mean 'before the present reign', and refers to the summer of 1483 when the two Queen Mothers, one past and one to be, planned the marriage of their children Henry Tudor and Elizabeth of York.[2] Elizabeth herself was the damsel constant to the man she had 'once promised and agreed to', and Henry was the plighted husband, who had indeed 'put his life oft in jeopardy' for her. On 19 September 1486, almost over-promptly,[3] Elizabeth had produced Henry's heir, a second Arthur (as Caxton's acquaintances Carmelianus and Gigliis pointed

---

in 1487 masqueraded as the imprisoned son of Clarence!); and soon after Clarence's execution in 1478 Oxford was nearly drowned in the moat at Hammes while trying to escape, doubtless because he knew Clarence would have left revelations for which his own head might be forfeit.

[1] As the unique surviving copy of *Aymon* in John Rylands Library Manchester lacks the first and last leaves, Caxton's prologue to Oxford is known only from William Copland's reprint, 6 May 1554, of a lost Wynkyn de Worde edition of 8 May 1504.

[2] Caxton must have sold the manuscript to Margaret Beaufort through her envoy Lewis, who came to negotiate with Elizabeth Woodville in the sanctuary of the Abbey, so near his shop. But Margaret herself, whom Richard had summoned to Westminster to bear his wife's train at his coronation on 6 July 1483, may well have visited her in person and even have met her ally the Printer.

[3] Exactly eight months after their marriage on 18 January 1486, so that Henry has not unreasonably been suspected of anticipating the ceremony in order to make sure that his queen was capable of childbearing.

out in their Latin nativity poems)[1] at Arthurian Winchester, the English Camelot, with Oxford as godfather and her mother Elizabeth Woodville as godmother. So, as Caxton pleasantly implies, Lady Margaret had saved the realm, not unaided by the example of the book he had sold her in perilous times, on the 'great adventures' of Blanchardin prince of Frisia and Eglantine called 'the Proud in Love', queen of Tormaday, 'tofore they might attain for to come to the final conclusion of their desired love'.

Early in 1489 Caxton unexpectedly produced a third edition of *Dicts of the Philosophers*, with its praise of Anthony Earl Rivers unchanged, no doubt under commission from the Dowager Queen his sister, in order to improve the Woodville family image and to counterbalance her exclusion from the credit titling of *Blanchardin and Eglantine*. Queen Elizabeth Woodville had kept up her refuge in the Abbey, with its security, gentility, piety, and proximity to Caxton and the Court, by taking a lease of the house called Cheyneygates within the precincts, which Abbot Eastney and the Chapter granted her on 10 July 1486 for a mere £10 yearly as their 'very especial good lord'. She had no doubt rewarded Caxton for inserting in his *Golden Legend* reprint of 1487-8 a brief life of St. Erasmus, the patron saint of childbirth, whose chapel she had founded in the Abbey in 1479 in memory of the birth there of her son Edward the little Prince in 1470. Caxton's allusion would point out discreetly her claim as the King's mother-in-law to equal credit with the King's mother in arranging within the Sanctuary of that Abbey the dynastic marriage of the Two Roses, her importance as mother of the child-bearing Queen and grandmother of the newborn Arthur. But Henry had never forgiven her for surrendering his betrothed to Richard in March 1484, and she was now in disgrace again for suspected complicity in the Yorkist rebellion of Lambert Simnel. At the council held at Sheen in February 1487 to deal with the Simnel menace Henry had confiscated her dower properties and put her away in Bermondsey Abbey (often used as a retirement home for unwanted female royalties) on a smallish pension. While he was about it he ordered Oxford to arrest her slippery son Dorset and kept him in the Tower till after Stoke. Her brother Sir Edward Woodville had taken a private army to Brittany to fight the French, ostensibly against Henry's wishes but evidently hoping to please him, and was killed with all his men at St. Aubin-du-Cormier on 28 July 1488. She was let out from Bermondsey now and then for minor royal occasions. For instance, she was there at the embarrassing moment on 9 May 1489 when the Pope's commissioner Perseus de Malviciis opened the court collecting-box for his master's Jubilee and Crusade in the presence of 'the King, Queen, and their mothers', and found it to contain only £11 11s; and in November 1489 the French ambassador

---

[1] Carmelianus wrote: *Arthurus rediit per saecula tanta sepultus . . . Arthurum quisquis praedixerat esse secundo Venturum . . .*' and Caxton must surely have done a roaring trade with his *Morte d'Arthur*.

was given audience by the newly pregnant Queen with 'her mother Queen Elizabeth and my Lady the King's Mother'. But her influence was gone for ever. So Caxton's *Dicts III* was his last book for the Woodville connection. It has not previously been dated; but a third break in the frameline of Caxton's device, 10 mm. further leftward and measuring 6 mm., a state found only here, shows that it was printed after *Reynard II* and *Directorium II*, but before *Doctrinal of Sapience*.

Another commission early in 1489 came from Caxton's old acquaintance Johannes de Gigliis. Gigliis was in high favour with Henry VII, having reported enthusiastically to Innocent VIII on his accession, and helped to secure the Pope's approval for his marriage to a fourth cousin.[1] Gigliis as papal collector in England, together with his colleague Perseus de Malviciis (whom Innocent had just sent with a sword of honour for Henry to induce him to join the Crusade),[2] ordered a new *Indulgence* for war against the Turks. This 1489 *Indulgence* granted, in return for 'only four, three, two or one gold florin', no more than the usual permission to choose one's own confessor once in a lifetime and again at the point of death. Caxton printed two editions, or rather impressed two different settings simultaneously on each folio leaf which were then cut up for separate sale, using his small type 7: 84G. (the reduced version of his type 5 supplied in 1485–6 by Veldener) for the first and only time.

The *Indulgence* was already on sale by 24 April 1489, when a copy now in the British Library was issued to a 'Master Henricus Bost', who has not previously been identified. This was none other than Henry Bost, Provost of Eton College from 31 March 1477 to February 1504. Perhaps Caxton knew him through Earl Rivers and Queen Elizabeth Woodville, for whom the Provost instituted early morning prayers in College Chapel (said to have been continued until the reign of Edward VI) in gratitude for their benefactions. Bost also had the interesting job of confessor to Edward IV's favourite mistress Jane Shore, whose two adorable portraits (one halflength and undressed for the bath, so that she may be said to be both topless and bottomless) the College still treasures. Edward was given to boasting that he had possessed 'the merriest, the wiliest, and the holiest harlot in his realm', Jane Shore being the first of these three, and Sir Thomas More wrote: 'Many he had, but her he loved.'[3]

---

[1] Gigliis was born at Bruges in 1434 (did Caxton know him as a boy?), took his doctorate of canon law at Oxford, received English citizenship in 1477, became canon of Wells in 1478 and prebendary of Lincoln in 1484, resident English ambassador in Rome later in 1489 and Bishop of Worcester (in absence, with his nephew Silvester as proxy) in 1497, and died still in Rome on 25 August 1498.

[2] Henry duly proclaimed the Crusade in 1490, but nothing happened. Sixtus had made the same gift of a sword to Edward IV in 1481, with the same result.

[3] After Edward's death in April 1483 Jane Shore was successively protected by Dorset, Hastings, and Richard III's solicitor Thomas Lynam. Richard

Sixtus IV had obliged Edward in 1476 by annulling Jane's marriage to the mercer William Shore on the ground of impotence, while Edward showed his appreciation on 4 December 1476 by granting royal protection to the complaisant husband. Caxton must have known his fellow mercer Shore very well, for on 7 May 1487, in one of those fictitious transactions for avoidance of debt or confiscation to which we have seen Caxton himself resorting in 1453, Shore made over his entire property to three friends including Caxton the Printer.[1] Another possible acquaintance of Caxton who bought this *Indulgence* was John Pamping, the employee of the Paston family who had disgraced himself in 1473 by wanting to marry their daughter Anne. I have seen an unrecorded copy in private hands which was made out to him on 3 March 1490.

Meanwhile, using the new patronage of Oxford and Lady Margaret the Queen Mother, Caxton reached his goal in the King himself. His book for Henry was *Faytes of Arms*, a treatise on the art and rules of war compiled in 1408-9, mostly from the fourth-century Roman Vegetius's *De re militari* and the thirteenth-century Honoré Bonet's *Arbre des batailles*, by the literary lady Christine de Pisan, the same who wrote the *Moral Proverbs* printed by Caxton for Earl Rivers in 1478. Christine, while admitting that the subject was 'thing not accustomed to women, which commonly do not enterprise but to spin on the distaff and occupy them in things of household', was briefed by her many soldier friends with uptodate technical information on gunnery, gunpowder, siegecraft and military ethics, so that her book remained of practical value in Caxton's time. The manuscript, says Caxton, 'being in French was delivered to me William Caxton by . . . my natural and sovereign lord King Henry the VII in his palace of Westminster' on 23 January 1489.[2] The King 'desired and willed me to translate this said book . . . and to emprint it to the end that every gentleman born to arms and all manner men of war, captains, soldiers, victuallers and all other should have knowledge how to behave them in the faytes of war and of battles, and so delivered me the said book then my lord the Earl of

---

charged her with witchcraft, made her do public penance as a harlot, imprisoned her in Ludgate, and confiscated her goods. Historians have absurdly believed that Dorset, Hastings and Lynam (who married her) were carried away by passion, and Richard by moral disapproval. Obviously the incentive of all four was the wealth Edward had given her.

[1] A statute invalidating such deeds of gift, and deploring that 'persons that maketh the said deed of gift goeth to sanctuary and occupieth and liveth with the said goods', was enacted by Parliament in November 1487. Abbot Eastney had been reprimanded by Henry's Council the year before for permitting such misuse of sanctuary. Was William Shore in Westminster Sanctuary next door to Caxton?

[2] The manuscript was not, as Blades suggests, British Library Royal 15.e.vi, for this omits Christine's anti-English version of the renewal of the Hundred Years' War in 1367 (alleging that the Black Prince murdered the French envoys, whereas the chivalrous Prince only threw them into prison), which is present in Caxton's text.

Oxford awaiting on his said grace'. 'I think my labour well employed,' says Caxton, reaching at last the propaganda message for which Henry has sponsored him, 'for to have the name to be one of the little servants to the highest and most Christian King and prince of the world, whom I beseech Almighty God to preserve, keep and continue in his noble and most redoubted enterprises as well in Brittany, Flanders and other places that he may have victory, honour and renown to his perpetual glory. For I have not heard nor read that any prince has subdued his subjects with less hurt et cetera, and also helped his neighbours and friends out of this land.' Caxton's words apply exactly to events of the spring and summer of 1489. In February and March Henry had made treaties of alliance against France with Brittany, Flanders and Spain; he sent an expeditionary force to Brittany in April and won the battle of Dixmude for Maximilian in Flanders on 13 June; and he put down rebels in Yorkshire, who had murdered the tax-gathering Earl of Northumberland on 28 April, by hanging the ringleaders and sparing the 'poor commons'. Caxton's propaganda was designed not only to glorify Henry, but to justify the unpopular taxation of that year by the King's successes.

Instead of setting to work immediately to complete this book received on 23 January Caxton first translated *Doctrinal of Sapience*, which he finished on 7 May 1489. He would hardly have procrastinated with the royal command for *Faytes of Arms* unless Oxford had instructed him to delay publication until the march of events made the time ripe for his political message. In the end he was obliged to go to press before finishing his translation, and completed it only six days ahead of his compositors. 'Which translation was finished the 8th day of July the said year and emprinted the 14th day of July next following and full finished', he boasts, making one of his humorous riddles on the mysteries of printing with the emphatic 'next following and full finished', rather as in *Recuyell* with the famous 'begun in one day and also finished in one day'.[1]

Doctrinal of Sapience, which no doubt appeared soon after *Faytes of Arms*, is a manual of popular theology with pious, facetious, ribald or horrific anecdotes, suitable both for lay devotional reading and as sermon material for preachers, and very similar to *Royal Book* and *Book of Good Manners*. Indeed, the anonymous author[2] borrowed part of his material word for word from *Somme le Roi*, but Caxton, whether unaware of this or scorning to repeat himself, translated the same French into different English. Caxton evidently worked under expert ecclesiastical supervision from his unnamed client, for he

[1] Blades and others have supposed that 'next following' means the year 1490; but the historical context and Caxton's little joke make it certain that the year is 1489.

[2] Caxton, or his manuscript, mistakenly ascribes the French original, written in 1388, to Guy de Roye (1345-1409), Archbishop of Sens, who was only the dedicatee. Blades wrongly identifies it with an entirely different work, the Latin *Manipulus curatorum* by Guido de Monte Rocherii.

werkes of mercy.and hath kepte all for his sone. And by the
comaundement of our lord J haue taken away the matere ƶ
the cause of his anaryce. And haue sente the soule of the childe
in to heuen whiche was innocent. And when the impte had
herde all this/he was delyuered of all the euyll tamptacyons
that he had.ƶ glorified god ƶ hys obscure iugementis whiche
ben as a grete so wolo we without bottome.lyke as sayth
the prophete Thou whiche haste herde the iugementis of god
put them in thyn herte:for to gyue to the example for to suffre
wel.and to endure alle tribulacions and alle euyllys pacyent
ly. For we haue also example that more and better auaylleth
ƶ ought to moeue vs than the exaple whiche we haue sayd ƶ
that is of our lord Jhesu crist whiche fro his glorie of heuene
descended in to the valeye of myserye that is in to this worlde
after that he was born of the virgyne marye vnto the time that
he was put on the tree of the crosse he had alleway pouerte.pey

ne.reproches.and tribula
cyons:And in thende suf
fred deth and ryght dolo-
rouse passiõ as thou shalt
here now recyted wyth
the helpe of hym. whiche
gyue vs grace amen/
⁂ Of the passyon of our
lord Capitulo   vj

Now late vs behol
de by pitie. how the
sone of god wold suffre
for vs.how wel that we
may not wryte all.Ne
uertheles some thyng we
shal be tolde to thende that we may loue hym the better ƶ for

B j

omitted a section on the various possible mishaps which might invalidate a mass and the rules for dealing with these, but printed a special issue (represented only by a unique copy on vellum now in the Royal Library at Windsor) with an appendix containing this matter, concluding with the remark: 'This chapter tofore I durst not set in the book by cause it is not convenient nor appertaining that every lay man should know it.' Rather surprisingly *Doctrinal of Sapience* was printed in the gothic type 5, instead of the bâtarde type 6 in which Caxton produced all his other new English texts from 1488 onwards; perhaps he wished to make it uniform in type with the analogous *Book of Good Manners* and *Royal Book*. Like *Royal Book*, *Doctrinal of Sapience* is illustrated with woodcuts from the *Speculum vitae Christi* set.

A second edition of *Mirror of the World* must have appeared after *Doctrinal* and towards the end of 1489, as both books contain Caxton's device in a fourth state, in which the three frameline gaps have run into one, measuring 44 mm. *Mirror II* is illustrated with the original woodcuts used in *Mirror I* in 1481, except one of God the Father creating the world, which was no doubt lost or damaged, as it is replaced quite inappropriately by a cut of the Transfiguration from the *Speculum* set. Perhaps the second edition (though Caxton did not trouble to alter his prologue or epilogue) was subsidised like the first by the goldsmith Hugh Bryce, who was prospering more than ever under Henry Tudor. Bryce was elected Lord Mayor of London in October 1485, knighted by Henry in January 1486, and kept under Lord Giles Daubeney the post of Governor of the Mint which he had held under Hastings, until his death on 22 September 1496.

So, during 1489, Caxton had returned to Court favour and full production, but had less than two more years to live.

17

THE ART OF DYING

THANKS to his new connection with the Crown Caxton was entrusted with printing *Statutes 1, 3, 4 Henry VII*, the statutes of Henry's first three Parliaments held in 1485, 1487, and 1489-90. This book has always been assigned to the autumn of 1489, on the mistaken assumption that it ends with the close of Henry's fourth regnal year on 21 August 1489; but in fact it includes the final acts of the third Parliament, which were passed in its last session of 25 January-27 February 1490, and must therefore have been printed after that date, probably in March or April 1490. Several of the acts Caxton printed touched him nearly. 'Revocation of King Richard's act against Italians' (1 Henry VII cap. 10) repealed Richard's restrictions against foreign merchants, but automatically retained the saving clause permitting the import of foreign books, of which Caxton had taken advantage in his dealings with Guillaume Maynyal in 1487-8. 'Deeds of gifts of goods to the use of the maker of such gifts be void' (3 Henry VII cap. 4) annulled his deal with Jane Shore's ex-husband William soon after it was made. The third Parliament had opened on 13 January 1489 to grant the unpopular war taxation for which Caxton had made propaganda in *Faytes of Arms*. Hitherto the Statutes of England had always appeared in the curious jargon of law French, but now and ever after the printed language was plain English. No doubt the Tudor government was content to make its laws intelligible not only to the closed circle of lawyers but to the country justices and common men whom they most affected. But Caxton's own circumstances may have been the immediate cause of this epochmaking reform. William de Machlinia, who had printed Richard's *Statutes* of 1484 in the usual law French, was no longer in business, and Caxton not only lacked the specialist knowledge of law French, but possessed no type with the repertoire of special contractions which was considered necessary for such texts whether in manuscript or in print. Caxton was now the only printer in England; so, if Henry's laws were to be printed at all, they must be in English.[1]

In 1490 Caxton reached the peak of his reconquest of royal favour by presenting his *Eneydos* to Arthur Prince of Wales, 'my tocoming natural and sovereign lord', using the same titles as when long ago in 1477 he had dedicated *Jason* to the illfated little Prince Edward.

[1] After Caxton's death Wynkyn de Worde kept the monopoly of the *Statutes*, which he continued to print in English, while Pynson, who was trained as a specialist in law printing, produced the annual law reports or *Year Books*, and other works for which law French was still compulsory.

His French original, *Livre des Énéides*,[1] was a late fourteenth-century chivalric romance adapted from Virgil and Boccaccio's *Fall of Princes*. 'After diverse works made, translated and achieved, having no work in hand,' Caxton tells us, 'I sitting in my study whereas lay many diverse pamphlets and books, happened that to my hand came a little book in French . . . in which book I had great pleasure, by cause of the fair and honest words in French, which I never saw tofore like nor none so pleasant nor so well ordered. I delibered and concluded to translate it into English and forthwith took a pen and ink and wrote a leaf or twain, which I oversaw again to correct it. And when I saw the fair and strange terms therein I doubted that it should not please some gentlemen which late blamed me, saying that in my translations I had over curious terms which could not be understood of common people, and desired me to use old and homely terms in my translations.' These perspicacious gentlemen had detected Caxton in a lifelong mockmodest pretence which has often deceived his modern critics! Ever since *Recuyell* he had invariably apologised in his prologues for his 'rude and simple translation'. 'There be no gay terms nor subtle nor new eloquence,' he confessed in *Charles the Great*, 'no curious nor gay terms of rhetoric,' as he put it in *Faytes of Arms*, and in *Blanchardin and Eglantine* he beseeched Lady Margaret 'to pardon me of the rude and common English, for I confess me not learned nor knowing the art of rhetoric nor of such gay terms as now be said in these days and used'. But the 'gentlemen' were aware that Caxton's disclaimers were a mere 'humility formula', and that his style of translation tended, especially in romances of chivalry, to be 'curious' and 'gay' to excess. 'Fain would I satisfy every man,' Caxton proceeds, 'and so to do took an old book and read therein, and certainly the English was so rude and broad that I could not well understand it. And also my Lord Abbot of Westminster did do show me late certain evidences written in old English for to reduce it into our English now used. And certainly it was written in such wise that it was more like to Dutch than to English, I could not reduce it nor bring it to be understood.[2] And certainly our language now used varyeth far from that which was used and spoken when I was born, for we Englishmen be born under the domination of the moon, which is never steadfast but ever wavering, waxing one season and waneth and decreaseth another season. And that common English that is spoken in one shire varyeth from another. Insomuch that in my days,' says Caxton, beginning to tell

---

[1] First printed by Guillaume Le Roy, Lyons, 30 September 1483; but many textual differences show that Caxton cannot have used Le Roy's edition, and must as usual have translated from a manuscript.

[2] Abbot Eastney doubtless showed Caxton some of the Anglosaxon charters (an 'evidence' meant a legal document, especially a title deed), held by Westminster Abbey before the Dissolution, such as the one granted by Edward the Confessor, which is reproduced in Anglosaxon type by Stow, or those from King Edgar (A.D. 959) or Ethelred (about 1010 in a copy of about 1306), which still survive.

fayn wolde I satyſfye euery man/ and ſo to doo toke an olde
boke and redde therin/and certaynly the englyſſhe was ſo ru
de and brood that I coude not wele vnderſtande it. And alſo
my lorde abbot of weſtmynſter ded do ſhewe to me late certa:
yn euydences wryton in olde englyſſhe for to reduce it in to
our englyſſhe now vſid/ And certaynly it was wreton in
ſuche wyſe that it was more lyke to dutche than englyſſhe
I coude not reduce ne brynge it to be vnderſtonden/ And cer:
taynly our langage now vſed varyeth ferre from that. Whi
che was vſed and ſpoken whan I was borne/ For we en:
glyſſhe men/ben borne vnder the dompnacyon of the mone.
Whiche is neuer ſtedfaſte/But euer wauerynge/weyynge o:
ne ſeaſon/ and waneth & dyſcreaſeth another ſeaſon/ And
that commyn englyſſhe that is ſpoken in one ſhyre varyeth
from a nother. In ſo moche that in my dayes happened that
certayn marchautes were in a ſhip in tamyſe for to haue
ſayled ouer the ſee into zelande/and for lacke of wynde thei
taryed atte forlond.and wente to lande for to refreſhe them
And one of theym named ſheffelde a mercer cam in to an
hows and axed for mete .and ſpecyally he axyd after eggys
And the goode wyf anſwerde.that ſhe coude ſpeke no fren:
ſhe . And the marchaut was angry.for he alſo coude ſpeke
no frenſhe. But wolde haue hadde egges/ and ſhe vnderſtode
hym not/ And thenne at laſte a nother ſayd that he wolde
haue eyren/then the good wyf ſayd that ſhe vnderſtod hym
wel/Loo what ſholde a man in thyſe dayes now wryte.eg:
ges or eyren/ certaynly it is harde to playſe euery man/ by
cauſe of dyuerſite & chauge of langage . For in theſe dayes
euery man that is in ony reputacyon in his coutre. wyll vt
ter his compynycacyon and maters in ſuche maners & ter:
mes/that fewe men ſhall vnderſtonde theym/ And ſom ho:

7 Caxton's Type 6, from Eneydos, 1490.

his most celebrated story, 'happened that certain merchants were in a ship in Thames for to have sailed over the sea into Zeeland. And for lack of wind they tarried at Foreland and went to land for to refresh them, and one of them named Sheffelde a mercer came into an house and asked for meat and specially he asked after eggs. And the good wife answered that she could speak no French, and the merchant was angry for he also could speak no French, but would have had eggs, and she understood him not. And then at last another said that he would have *eyren*; then the good wife said that she understood him well. Lo, what should a man in these days now write, *eggs* or *eyren*?[1] Certainly it is hard to please every man by cause of diversity and change of language . . . And some honest and great clerks have been with me and desired me to write the most curious terms that I could find, and thus between plain rude and curious I stand abashed. But in my judgment the common terms that be daily used be lighter to be understood than the old and ancient English. And forasmuch as this present book is not for a rude uplandish man to labour therein nor read it, but only for a clerk and a noble gentleman that feeleth and understandeth in faytes of arms, in love and in noble chivalry, therefore in a mean between both I have reduced and translated this said book into our English not over rude nor curious, but in such terms as shall be understood by God's grace according to my copy. And if any man will engage in reading of it and findeth such terms that he cannot understand, let him go read and learn Virgil or the Epistles of Ovid,[2] and there he shall see and understand lightly all if he have a good reader and informer.'

Lastly Caxton appealed to 'Master John Skelton, late created poet laureate in the University of Oxford to oversee and correct this said book . . . For him I know for sufficient to expound and english every difficulty that is therein, for he hath late translated the Epistles of Tully[3] and the book of Diodorus Siculus and diverse other works out of Latin into English, not in rude and old language but in polished and ornate terms craftily, as he that hath read Virgil, Ovid, Tully and

[1] Caxton's colleagues were no doubt bound for Middelburg in Zeeland, and waited for their wind at Margate, the nearest anchorage within the Thames estuary to North Foreland. Blake has identified Sheffelde with a mercer of that name who issued from apprenticeship with Robert Hallom in 1456-7. Part of Caxton's joke is that these language difficulties occurred in his native Kent, where, as he had said in *Recuyell*, 'I doubt not is spoken as broad and rude English as is in any place of England'.

[2] Caxton (as he does in his prologues more often than editors have noticed) uses words from the book he has just translated. The author of *Livre des Énéides* omits the whole Sixth Book of the *Aeneid*, remarking merely: 'Who that will know how Aeneas went to hell, let him read Virgil, Claudian or the Epistles of Ovid and there he shall find more than truth'. He refers to Claudian's *De raptu Proserpinae* and Ovid's epistle from Dido to Aeneas (*Heroides* vii), although these contain nothing about Aeneas in Hades, and Caxton takes him on trust.

[3] Cicero, *Epistolae ad familiares*, a lost work which Skelton himself mentions in the list of his writings in *Garland of Laurel*: 'Of Tully's Familiars the translation'.

all the other noble poets and orators to me unknown. And also he hath read the nine muses and understands their musical sciences and to whom of them each science is appropriated, I suppose he hath drunken of Helicon's well.' Caxton's allusion is to Skelton's discussion of the Muses' functions in his translation of Poggio's Latin version of Diodorus Siculus (including 'Dame Erato which through her bounteous promotion her scholars and disciples bringeth unto so noble advancement'), and his plea to Clio:

> *'Mine homely rudeness and dryness to expel*
> *With the fresh waters of Helicon's well'*

in the lament which Skelton had written for the taxgathering Earl of Northumberland soon after his murder in April 1489. Indeed, the theme of Caxton's whole prologue evidently springs from the poet's words 'homely rudeness', and Skelton must have been chief among the 'great clerks' who had urged him 'to write the most curious terms that I could find.'[1]

Caxton's friendly banter with the poet is clearly connected in some way with the dedication of *Eneydos* to Prince Arthur, for whose creation as Prince of Wales on 29 November 1489 Skelton had written a poem which no longer survives. Skelton can hardly have been the Prince's tutor already, as some have supposed, for in the summer of 1490 Arthur was still a baby less than four years old. Indeed, there seems to be no positive evidence either that Skelton ever held the post or that he did not. But Caxton's testimonial suggests that he was at least either designate or a candidate for it. However this may be, after becoming (beyond all doubt) tutor to Henry's second son the future Henry VIII (born 28 June 1491) towards 1498, Skelton remembered Caxton's prologue and wrote:

> *'The honour of England I learned to spell . . .*
> *I gave him drink of the sugared well*
> *Of Helicon's waters crystalline,*
> *Acquainting him with the Muses nine.'*

Caxton completed the translation of *Eneydos* on 22 June 1490, and no doubt printed the book soon after. It contains his device in a fifth state, with the gap measuring 70 mm., but showing inked traces of the damaged frameline. The same state also appears in a second edition of *Speculum vitae Christi*, which can therefore be assigned to the same year, either before or after *Eneydos*. *Speculum II* is a page for page reprint of *Speculum I*, using the same gothic type 5 and the same woodcut illustrations, so that casting off was not required. This motive of convenience, as well as reluctance to waste paper, explains why Caxton did not change to the more suitable type 6 for this

---

[1] Blades and Duff are mistaken in saying that Skelton actually revised *Eneydos* for press. Caxton only invites him in jest to improve it after publication.

English text.[1] The two editions can be distinguished at a glance from the spelling of the chapter-headings, which read 'Ca.' throughout in *Speculum I*, whereas in *Speculum II* they vary in the first half of the book but read consistently 'Capitulum' in the second half (quires k–s), a sure sign that the book was printed on two presses.[2]

Three other works in which Caxton used type 5 for its original purpose as a Latin liturgical type probably belong also to 1490. An octavo fourth edition of *Horae ad usum Sarum*, known only in two fragments of four leaves each, is apparently a page-for-page reprint of *Horae III* but differs by containing red printing. Similar red printing occurs in the quarto *Officium transfigurationis Jesu Christi*, giving the service for the Feast of the Transfiguration on 6 August instituted by Pope Calixtus III to celebrate the victory of Hunyadi Janos over the Turks in Hungary on that day in 1457, and intended for insertion in manuscript breviaries to bring them up to date. Caxton illustrated this with a small woodcut of Christ transfigured (as in *Matthew* xvii) between Moses and Elijah, with the disciples Peter, James and John at his feet, evidently by the quaint Flemish artist of the *Horae* series. A companion piece is *Commemoratio lamentationis B.V.M.*, giving the still more recent office variously known as the Lamentation, Compassion, or Pity (Caxton names all three alternatives), and later as the Seven Sorrows of the Virgin Mary, celebrated on the Friday before Passion Sunday, and probably promulgated in this form by Sixtus IV towards 1475 (like the Feast of the Visitation which Caxton had printed in 1480) as propaganda for his new church of Santa Maria del Popolo in Rome. Both pieces (and this is a good reason for placing them together and in 1490) are signed only on the first and third leaves of the quire, a method which implies quarto printing by the whole sheet, that is with a forme containing four pages. This pattern of signatures occurs only here in Caxton's work, though it became normal in Wynkyn de Worde's quartos. In 1491 Caxton continued to print his quartos by the whole sheet, but reverted to his earlier practice of signing them by the half sheet.

Entries in the Exchequer accounts for the 'term' between Easter 5 Henry VII (11 April 1490) and Michaelmas 6 Henry VII (30 September 1490)[3] show three small payments to Caxton, totalling

---

[1] Owing to its larger size type 6 required more paper for the same text than type 5. Perhaps Caxton remembered his unfortunate experience in *Directorium sacerdotum II*, when the change of type had compelled him to use 20 per cent more paper than in the first edition.

[2] In a copy of *Speculum II* in the National Library of Wales at Aberystwyth the contents of the inner forme of the third sheet in quire e (e3 verso–e6 recto) appear in reverse order, a common accident in Caxton's work, which he rectified after detection by reprinting the whole sheet. Blades interpreted this mishap as proving that Caxton was still printing a page at a time, whereas in fact it shows just the opposite, that he was printing by the forme, as he always did from 1480 onwards. The British Library *Speculum II* is on vellum.

[3] Crotch wrongly dates these entries as '1490–91', forgetting that Henry's regnal year began on 22 August and did not commence a new calendar year.

10s 10d, 'for the expenses of diverse officers of the King's Receipt at Westminster attending for certain matters for the King'. The Prior's Chamber at the Abbey was sometimes borrowed by the King's Receipt. Probably, as the recorded sums are so trivial, Caxton had merely obliged these Treasury officials with stationery or temporary extra accommodation in his office. But the last entry, for 3 July 1490, specifies that the officials were 'attending there for diverse appointments to be made for the Sea and otherwise', and it is tempting to guess that the aged Caxton may have been called in once more as a shipping expert in the intense diplomatic activity of that summer, when Henry's envoys to the Pope's peace mission at Calais and the Duchess Anne in Brittany were moving to and fro across the Channel.

At the same time Caxton apparently planned an edition of Mandeville's *Travels*, one of the most delightful and unreliable books of the middle ages, perhaps on commission for St. Albans Abbey. The original binding of British Library MS. Egerton 1982, which is second in textual importance only to the Cottonian among English Mandeville manuscripts, contained a transcript made in 1579 by the Elizabethan publisher Richard Tottel of a still earlier inscription reading: 'This fair book I have from the Abbey at Saint Albans in this year of Our Lord 1490 the sixth day of April. William Caxton.' Sir John Mandeville, or the author masquerading under his name who died at Liège on 17 November 1372, claimed Saint Albans as his birthplace, and the Abbey, still more outrageously, was claimed as his place of burial, instead of the Williamites' church at Liège, where his genuine tomb lay. An early inscription to this effect remains in the cathedral to this day, on the second pillar north of the west door, and Caxton's acquaintance the St. Albans Schoolmaster Printer made a point of inserting a mention of this alleged tomb in the enlarged edition of Caxton's *Chronicles of England* which he compiled in 1483 and printed towards 1485. Perhaps Caxton was prevented from printing *Mandeville* by his death in 1491; but probably the plan was made inopportune by serious trouble between St. Albans and the Crown soon after he received the manuscript. On 5 July 1490 Chancellor Morton admonished Abbot William Wallingford on charges including theft of the Abbey treasures, and allowing his monks to 'resort as if to a brothel' to the nearby nunnery of Pré, where he had appointed a married woman friend named Eleanor Jermyn as prioress. In the end the first English *Mandeville* was printed by Pynson towards 1496 from an inferior abridged manuscript, reprinted by Wynkyn de Worde in 1499, and the unique Egerton text remained unpublished until 1889.

The churchwardens' biennial accounts of St. Margaret's Westminster from May 1490 to May 1492 include an entry in 1490 reading: 'Item at burying of Maude Caxton for torches and tapers 3s 2d'. Other evidence is lacking, and this Maude may, for example, have been only a widow, daughter-in-law or other relative of one of the other Westminster Caxtons such as the Oliver and William who were buried at the same church in 1465 and 1478 respectively.[1] Even so, Blades's suggestion that Maude Caxton was the wife of William Caxton the Printer may well be correct. Perhaps it is even no coincidence, but as Blades conjectured a response to Maude's death, that Caxton interrupted his work on *Eneydos* to translate the *Art of Dying*, which he completed on 15 June 1490, just a week before finishing *Eneydos* on 22 June.[2]

Caxton translated *Art to know well to die*, as he entitled it, from an abridged French version of the Latin *Ars moriendi*, a compilation of meditations, exhortations, prayers and maxims for the dying, their relatives, and ministering priests. Several different works went under the same name, and these were expanded, abridged and conflated into an unmapped labyrinth of varying texts containing more or less material in common. Colard Mansion had printed a different French version towards 1480, and Caxton himself was to print yet another English version in 1491. This lugubriously popular work fits into Caxton's continuous output of devotional books since 1486, and probably, like the rest, it was commissioned; but Caxton had a way of obtaining commissions for texts of his own choice, and it would be inhuman to imagine the old man twice printing an *Art of Dying* without thought of death, perhaps Maude's, but certainly his own.

Soon after this book on the art of dying Caxton produced one on how to keep alive. *Governal of Health* is sometimes regarded as the first printed English medical book; but this priority belongs to *Treatise on the Pestilence*, a textbook on bubonic plague printed by William de Machlinia towards 1485–6 in three editions (one containing the first English typeset titlepage)[3] to meet the demand caused by an entirely different epidemic, the sweating sickness. In fact *Governal of Health*, an anonymous early fifteenth-century translation from one of the many versions of the Latin *Regimen sanitatis*, is a manual of everyday hygiene rather than medicine, full of admirably practical advice on diet, exercise and sleep. One section sheds a lurid light on

[1] Maude Caxton must have been a person of some importance, for her torch and taper expenses are higher than most. Oliver Caxton's cost only 8d, the other William Caxton's 20d, and the usual fee was from twopence to sixpence.

[2] Blades errs in suggesting that Maude's death would 'explain, in the most interesting manner, the reason why he in that year (1490) suspended printing *Faytes of Arms* until he had finished a new undertaking, *The Art to die well*', not only because the true date of *Faytes of Arms* is 14 July 1489, not 1490, but because for Caxton translation and printing (which he did not perform in person) were concurrent and compatible activities. It was the translation of *Eneydos* that overlapped with *Art to know well to die*. However, such clashes were not unprecedented, for Caxton had worked concurrently on *Godfrey of Boloyne* (begun 12 March, finished 7 June 1481) and *Reynard the Fox* (finished 6 June 1481), and probably also on *Faytes* and *Doctrinal* in 1489.

[3] However, Caxton's fullpage frontispiece to Aesop's *Fable's* 26 March 1484, which his woodcutter copied from the Lyons 1480 edition, includes the lettering ESOPUS over a portrait of Aesop, and might be claimed as a woodcut titlepage.

¶ Here begynneth a lytyll treatyse schortely
compyled and called ars moriendi/that is
to saye the craft for to deye for the helthe of
mannes sowle.

When ony of lyklyhode shal deye/thenne
is moste necessarye to haue a specyall
frende/the whiche wyll hertly helpe and praye
for hym & therwyth counseyll the syke for the
wele of his sowle/& more ouer to see that alle
other so do aboute hym/or ellys qupckly for to
make hem departe. ¶ Thenne is to be remem
bred the grete benefytes of god done for hym
vnto that tyme/and specyally of ye passyon of
our lorde/and thenne is to be rede somme story
of sayntes or the vij psalmes wyth ye letanye
or our lady psalter in parte or hole wyth other
And euer the ymage of the crucyfyxe is to be
hadde in his syght wyth other. And holy wa
ter is oftymes to be cast vpon and about hym
for auoydyng of euyll spirytes ye whiche they be
be full redy to take theyr auauntage of the
sowle yf they may. ¶ And thenne and euer
make hym crye for mercy and grace & for the

A i

VI Caxton's Type 8, from Ars Moriendi, 1491.

the gymnastic home life of 'great prelates', for whom the outdoor sports recommended 'for young men that be lusty, to run, to wrestle, to leap, to cast the stone' would be too undignified: 'For which in chamber shall be a great cord knitted in the end and hanged up, and take that cord with both hands and stand upright so that thou touch not the earth and stand a good while, then run as much as thou mayst hither and thither with that cord and otherwise skip . . . or thus, close a penny in thine hand and let another take it if he may.' 'Sleep first on the right side,' says the prescription against insomnia, 'for thy digestion shall be better, for then lieth thy liver under the stomach as fire under a cauldron, and after thy first sleep turn on thy left side that thy right side may be rested, and when thou hast lain thereon a good while and slept, turn again on thy right side and sleep all night forth.' Caxton (or his manuscript) added an abridged version of *Medicina stomachi*, or *Dietary*, attributed in British Library MS. Harleian 116 to 'the monk of Bury', i.e. Lydgate, and accepted by modern editors as a genuine part of the Lydgate canon, with a charming conclusion:

> *Moderate food giveth to man his health . . .*
> *And charity to the soul is due,*
> *This receipt bought is of no apothecary,*
> *Of Master Anthony nor of Master Hugh,*
> *To all indifferent it is richest dietary.*

Governal of Health probably belongs to 1491, because like other quartos of that year its signatures are printed in capital letters, a method which suggests that these were printed at the same time and intended for sale either separately or bound together according to the customer's wishes.

The remaining seven books of 1491, though none contains a date, can be assigned to that year with certainty because all contain either a new type, Caxton's last, or his device in its latest state of deterioration, or both.

Caxton's new type 8: 114.G is a largish gothic text and heading type in a characteristic Parisian style, of which almost identical founts were used in Paris from the later 1480s onwards by many wellknown printers, including Antoine Caillaut, Pierre Levet, Maynyal, Higman and Hopyl, all of whom no doubt acquired it from the same unidentified typefounder (perhaps the prototypographer Ulrich Gering, who himself soon afterwards used yet another variety of it). Probably Caxton obtained his supply through Maynyal, who had printed the *Sarum Missal* and *Legenda* for him in his own slightly different version of the same type.[1] Caxton's previous seven types had all been Flemish in style and presumably made by Veldener. This first recourse to French sources was no doubt in some degree an

---

[1] Maynyal 110.G is distinguishable from Caxton type 8 by its height, which is 4 mm. less, and by several different majuscules, including C, E, alternative 'Gallic' S, and T.

intentional modernising change of policy, which Caxton bequeathed to Wynkyn de Worde; for Wynkyn began by re-using Caxton's existing types 3, 4*, 6, 7 (as well as 8), but gradually replaced them with French types from Paris.[1] But Caxton was also at this stage guided by necessity, for he was in any case obliged to replace his overworked gothic type 5, and Veldener had gone out of business. Probably Caxton had not altogether severed his contacts with the Low Countries, for new woodcut initials from Gouda (where they had been used by the printer Gotfridus de Os in 1486) appear in Wynkyn de Worde's work immediately after Caxton's death, and were presumably ordered by Caxton himself when he was still alive.[2] Caxton used type 8 in only two surviving books, *Ars moriendi* and Mirk, *Festial,* second edition. It is not present (as Duff alleged, making two of his rare typographical errors) in Caxton's 1491 edition of *Quattuor sermones* nor in *Fifteen Oes,* both of which are entirely in type 6. However, as we shall see, *Fifteen Oes* can be shown, in a highly unusual way, to have been printed at a time when type 8 was simultaneously in use in two books now lost.

Further evidence on Caxton's last books is supplied by his vulnerable device, which shows further damage in a break in the left frameline near the foot. This state, the sixth and last in his lifetime, appears in three books: the above-mentioned *Festial II* and *Quattuor sermones III,* and *Book of Diverse Ghostly Matters.*

Book of Diverse Ghostly Matters is a three-in-one tract volume in which each of the three parts has its own set of printed signatures—in capitals (A-M, A-D) for the first two, and double letters (aa-cc) for the third part—in order to facilitate individual assemblage and binding for separate or combined sale. But the unity of the volume is shown by the collective title and contents list at the end: 'Thus endeth this present book composed of diverse fruitful ghostly matters' —ghostly, of course, meaning spiritual or devotional—'of which the foresaid names follow to the intent that well disposed persons that desire to hear or read ghostly informations may the sooner know by this little entitling the effects of this said little volume, in as much as the whole content of this little book is not of one matter only . . .' The three pieces are *Seven Points of True Love,* being an English version of the Latin *Horologium sapientiae* (i.e. *Dial of Wisdom*) written about 1340 by the Dominican Henricus Suso, the anonymous

---

[1] Rather curiously, Caxton's types 3 and 4* as used by Wynkyn (his types 6 and 1 respectively) were evidently recast. It seems easier to believe that these recasts were obtained from Veldener by Caxton in 1486 but laid aside and not used by him, than that they could still be made six and more years later by anyone else.

[2] This set, most of which are decorated with grotesque faces, was never used by Caxton, although Blades misleadingly included several letters from it (I, Q, and large H) in his plate of 'Woodcut initials from Caxton's books'. Caxton also apparently acquired but never used Gotfridus de Os's type 103G., which Wynkyn de Worde used once only, in his 1496 reprint of the St. Albans 1486 *Book of Hawking, Hunting, and Blasing of Arms.*

Twelve Profits of Tribulation (which was reprinted separately by Wynkyn de Worde in 1499), and an abridged *Rule of Saint Benedict*, 'for men and women of the habit thereof the which understand little Latin or none to the end that they may often read and execute the whole rule and the better keep it than it is', ending with the request: 'of your charity pray for the translator of this said treatise'. These items were neither translated nor compiled by Caxton, who evidently printed the collective manuscript volume as it reached him, 'by desiring', as he says, 'of certain worshipful persons'. These must surely have belonged to Caxton's Benedictine connection, whether at Westminster Abbey or elsewhere, and the book was perhaps intended to correct the unfortunate impression caused in 1490 by the disgrace of Abbot Sant at Abingdon for treason and of Abbot Wallingford at St. Albans for whoremongering.

Ars moriendi perhaps came last in Caxton's composite set of quarto tracts, for it is alone among these in containing type 8, which is here used (for incipit and headings) only in the outer sheet, as though it had just arrived and was still in short supply. The text is not (as some have said) a reprint of last year's *Art to know well to die*, but a different and still shorter version in only five leaves, followed by three more, containing a prayer for the dedication of a church and maxims on humility, obedience, patience and charity, which do not belong to *Ars moriendi* at all. This unique compilation must have had a longish evolution in English behind it, and the translation is not, as Blades asserts, 'doubtless by Caxton himself'. Even so Caxton must have read its recommendations with personal feeling: 'When any of likelihood shall die, then is most necessary to have a special friend the which will heartily help and pray for him and therewith counsel the sick for the weal of his soul . . .'

Caxton's second edition of *Festial* was reprinted not from his first edition of 30 June 1483, but from Rood's Oxford edition of 14 October 1486, which differs widely in wording and contents. Caxton's *Festial II*, however, includes for the first time an updating supplement, not found in Rood's edition, giving sermons for the new feasts of the Transfiguration and the Name of Jesus. No doubt this addition accounts for his need to use the Oxford edition, of which his client probably supplied him with a copy made up with the new material in manuscript. Caxton also copied Rood's edition by printing in double columns (which he had not done previously when using type 6), with an imposing height of 33 lines (instead of his usual 31), and with Latin words and quotations displayed in type 8 (Rood likewise had used a display type for these). *Quattuor sermones III* was reprinted without change from Caxton's second edition of 1484, as both these contain various correct readings replacing misprints in the *Quattuor sermones I* of 1483.

Caxton's last book for named patrons was a quarto edition of the prayerbook called in English *Fifteen Oes*, because each section begins with the exclamation 'O'. 'These prayers tofore written,' he says, 'be emprinted by the commandments of the most high and virtuous princess our liege lady Elizabeth by the grace of God Queen of England and of France, and also of the right high and most noble princess Margaret, mother unto our sovereign lord the King, by their most humble subject and servant William Caxton'. The Latin original was attributed to St. Bridget of Sweden (1302-73),[1] and soon became a popular component of Books of Hours, especially in England, where from the mid-fifteenth century onwards the *Sarum Horae* often included one of the several English translations[2] as well as the Latin text. Caxton's edition, though evidently intended for separate sale when desired, has all the appearance of being a supplement to a *Horae*, for it is decorated with new woodcut borders in *Horae* style, and contains a vivid and accomplished Crucifixion woodcut from the same series as four other cuts by the same Flemish artist which Wynkyn de Worde used in two *Sarum Horae* printed towards 1494. A strange piece of evidence shows that Caxton did in fact print a *Sarum Horae* as a companion to *Fifteen Oes*, although the book itself has not survived.

A copy of *Mirror of the World II* now in the British Library is one of a dozen surviving Caxton volumes, ranging in date from a *Jason* of 1477 to a 1491 *Festial II* with *Quattuor sermones III*, which still remain in contemporary bindings by the 'Caxton Binder'. This craftsman's work can be identified by its distinctive ornamental stamps, which are also found on various books not printed by Caxton, including Wynkyn de Worde's 1495 *Vitas patrum* and a Hagenau book as late as 1506, and several manuscripts associated with Westminster Abbey, including John Flete's history of the Abbey containing the scrap of press-waste from Caxton's Bruges *Recueil* of 1475-6.[3] The *Mirror II* binding contained a mutilated sheet of waste from *Fifteen Oes*, evidently discarded as unusable after one of Caxton's pressmens' awful mishaps.[4] Before this fragment was passed over to

[1] Modern scholars consider that the style is not hers, though the work was composed within her lifetime.

[2] The translation printed by Caxton is one of these, though Blades as usual attributes it to Caxton himself.

[3] It is not clear whether the Caxton Binder was an actual employee in Caxton's own establishment, or an independent craftsman working both for Caxton and for the Abbey. One wonders whether he may not have been the James Bookbinder who in 1499-1500 (and perhaps even from 1495-6) rented the shop next door to Caxton's former Chapter House premises (which Wynkyn de Worde continued to occupy from Caxton's death in 1491 until his move to London in 1500), and from 1505-6 to 1508-9 took over Caxton's and Wynkyn's former shop in addition. Many more examples of the Caxton Binder's work were no doubt destroyed by the collectors of the eighteenth and nineteenth centuries who saw fit to have their Caxtons rebound, for it occurs on a large proportion of those which still survive in contemporary bindings.

[4] This uncut inner sheet of the first quire (a3-a6)—giving visible evidence that Caxton was then printing quartos by the whole sheet, four pages at a time—has the inner forme imposed upside down, thus presenting the disastrous page sequence 3 recto, 5 verso, 6 recto, 4 verso, 5 recto, 3 verso, 4 recto, 6 verso! When the error was noticed the type for the outer forme had already been

Jhesu endeles swetnes of louyng soules / O Jhesu goostlie ioye passing & excedyng all gladnes and desires. O Jhesu helthe & tendre louer of al repentaunt sinners that likest to dwelle as thou saydest thy selfe with the children of men / For that was the cause why thou were incarnate and made man in the ende of the worlde. Haue mynde Blessed Jhesu of all the sorowes that thou suffredst in thy manhode drawynge nyhe to thy Blessed passion / In the whiche most holsom passion was ordeyned in the deuyne hert / By counseyle of al the hole trynyte . for the raunson of al mankynde. Haue mynde Blessed Jhesu of al the grete dredes & anguysshes & sorowes that thou suffredst in thy tendre flesshe afore thy passion on the crosse / whan thou there betraied of thy disciple Judas

8 Caxton's last book, Fifteen Oes, 1491.

the Caxton Binder it was used to dry off wet newly printed sheets from two other books which were simultaneously at press, for on one side it shows the blotted image or 'offset' of an eight-page octavo sheet with some red printing, and on the other side the offset of a four-page quarto sheet with each page framed in the woodcut borders used in *Fifteen Oes*. These impressions are not only faint and blurred, and partly obscured by being superimposed on the heavy black setting of *Fifteen Oes*, but are also of course inverted in mirror image so that they have to be examined in a mirror. Henry Bradshaw, the Cambridge University Librarian, the founder of modern study of fifteenth-century typography, and the too often disregarded mentor of William Blades, studied them in 1877 (when they were shown at the quatercentenary Caxton Exhibition at South Kensington), and suggested that the octavo offset derived from Caxton's fourth edition of *Horae ad usum Sarum*. The quarto offset, he thought, was an unknown work in Caxton's type 5. When I made my own study of Caxton through the Looking-Glass in 1961 it turned out that for once Bradshaw was mistaken, for both offsets are in type 8. The octavo offset is not from *Horae IV*, which was in type 5, but is the sole surviving relic of a lost fifth edition, an octavo *Horae V*. I was able to identify the Latin text of the quarto offset as coming from the Vigils of the Dead in the *Sarum Horae*; so this spectral image is all that remains of a lost *Horae VI*, printed along with *Horae V* soon after (or perhaps concurrently with) *Fifteen Oes*, and evidently intended as its companion.

Now that the Queen Dowager Elizabeth Woodville was put away at Bermondsey Abbey (where she died on 8 June 1492), the Queen her daughter was well understood to be under the rather strict charge of the King's mother. Caxton's dedication of *Fifteen Oes* is almost a political manifesto to this effect. Perhaps the royal ladies also commissioned the accompanying *Horae VI*, for Wynkyn de Worde repeated the same dedication to them both in his page-for-page reprint of the complete work towards 1494. Also in 1494 Wynkyn printed Walter Hylton, *Scala perfectionis*, again at Margaret Beaufort's request:

> *The King's mother of excellent bounty*
> *Henry the Seventh that Jesu him preserve*
> *This mighty princess hath commanded me*
> *To emprint this book her grace for to deserve . . .*

and with a loyal tribute to his dead master:

> *And Wynkyn de Worde this hath set in print*
> *In William Caxton's house so fell the case*
> *God rest his soul, in joy there mote it stint . . .*

---

distributed, so the sheet had to be reprinted with a new setting of this forme and with the still standing type of the inner forme (right side up this time).

A copy of this book, as Mr. P. J. Croft has shown, was presented jointly by Elizabeth and Margaret to Mary Roos, their lady in waiting. The royal ladies had evidently clubbed together and paid for part of the edition to give to their friends. Probably this procedure is typical of Caxton's whole system of patronage, as inherited by Wynkyn. Very likely most or all of his patrons—the royal Margaret of Burgundy, Clarence, Edward and Henry, their queens and children, the great lords Rivers, Hastings, Arundel, Oxford, the merchants or state officials Bryce, Pratt, Daubeney, the nameless 'diverse gentlemen' or 'worshipful persons'—had subsidised his books by undertaking (as he says of Arundel in *Golden Legend*) 'to take a reasonable quantity of them' for use as gifts. These bargains were clinched by the added inducement of his honorific prologues and epilogues, each with its message of praise for his clients' status and propaganda for their ambitions of the moment.

Meanwhile the indomitable old man was at work on one of his longest translations, *Vitas patrum*, a collection of lives of the Desert Fathers often wrongly ascribed to St. Jerome, for which he used the edition 'late translated out of Latin into French and diligently corrected in the city of Lyons the year of Our Lord 1486'.[1] Caxton must have begun soon after finishing *Eneydos* in June 1490, for this is an enormous work of about 350,000 words, second only to *Golden Legend* in length. He was now probably at least sixty-nine or seventy years old, yet there is no trace of diminishing output or vigour or relish in his last period. Caxton's complaint in *Recuyell* sixteen years before, that he had been forced to learn printing because 'my pen is worn, mine hand weary and not steadfast, mine eyes dimmed with overmuch looking on white paper, and my courage not so prone and ready to labour as it hath been, and age creepeth on me daily and feebleth all the body', can only have been intended, at that time, as one of his little jokes. Caxton had learned his art of dying by living, which for a writer means writing. He did not survive to see *Vitas patrum* in print. Wynkyn de Worde produced it in 1495, remarking that the text was 'translated out of French into English by William Caxton late dead and finished it at the last day of his life'.

The day and month of Caxton's death remain unknown, but four contemporary mentions unite to show that it came in 1491, probably towards the end of the year.

Wynkyn's own testimony seems to have gone unnoticed. *Vitas patrum*, he says, was 'in the year of Our Lord 1491 reduced into English', so Caxton's death in harness must have occurred in that year.

Caxton last paid his year's rent of 10s (it had remained unaltered since 1476, for in those golden days inflation ran at an annual rate of zero) to the Abbot-Sacrist John Eastney for his Chapter House shop at Michaelmas 1491; but when Eastney made up his annual accounts

[1] Printed by Nicolaus Philippi and Jean Dupré, 15 January '1486', i.e. 1487 by the modern calendar, at Lyons.

at Michaelmas 1492 Wynkyn de Worde had already taken over the tenancy.[1]

The biennial Churchwardens' accounts for St. Margaret's Westminster for the period from May 1490 to May 1492 contain an entry for Caxton's funeral, which was more than twice as expensive as Maude's: 'Item at burying of William Caxton for four torches 6s 8d, Item for the bell at same burying 6d'. This entry, Blades thought, can be dated from its position as 'towards the close of the year 1491'.

The eighteenth-century antiquary Joseph Ames recorded an inscription 'wrote down in a very old hand in a *Fructus temporum* of my friend Mr. Ballard's of Camden in Gloucestershire: 'Of your charity pray for the soul of Master William Caxton, that in his time was a man of much ornate and much renommed wisdom and cunning and deceased full Christianly the year of our Lord 1491:

> *Mother of Mercy shield him from the horrible fiend*
> *And bring him to life eternal that never hath end.'*

Although this book has never turned up since, there is no reason to doubt Ames's information. There is no such work as *Fructus temporum*: what Ames or Mr. Ballard of Chipping Campden meant was no doubt *Chronicles of England with the Fruit of Times*, an enlarged edition of Caxton's *Chronicles* with additions from Rolewinck's *Fasciculus temporum* which was compiled by the Schoolmaster Printer in 1483, printed by him towards 1485, reprinted by Wynkyn de Worde in 1497 and 1502, by Pynson in 1510, and by Julian Notary in 1504 and 1515. Duff arbitrarily identified Mr. Ballard's volume as Notary's edition of 1515, but it seems more likely that it was Wynkyn's of 1497, and that the writer was Wynkyn himself, or an employee or customer with a recent memory of Caxton's end.

However, Caxton's own output reveals the approximate time of his death, for none of his books can be later than 1491, and even those which are assignable to that year are hardly enough for a full twelve months' production; so a date of death towards the autumn of 1491 could be deduced even without confirmation from documentary evidence.

Caxton's will does not survive or has not been found, alas, for if it was as meticulously informative as his master Robert Large's it would reveal much that we do not know about his relatives, em-

[1] The entry reads: 'From William Caxton for the shop next door now in the possession of John Wynkyn, for a year, 10s.' Unfortunately, as Prior Robert Essex died in June 1491, the last entry in his account book for Caxton's tenancy of the Almonry premises (which came under the Prior) is for 31 January 1491, when 'W. Kaxton paid me for the three houses 23s 4d'. The account book of his successor Prior George Fascet (1491-8) does not survive. Wynkyn de Worde's regular imprint 'in Caxton's house' shows, however, that he continued to occupy Caxton's Almonry premises, and also, retrospectively, that Caxton himself had printed there. The Chapter House shop was no doubt a shop and nothing more.

ployees, friends, executors, and finances. He left copies of his *Sarum Legenda* to St. Margaret's church, as we have seen, probably to pay for 'obits' or annual services for his soul. Only one *Legenda* remained in 1502, and the last entry for such a gift appears in the church-wardens' account towards 1506: 'Item four printed books, two of them the *Life of St. Katherine* and other two of the *Birth of Our Lady* of the gift of the executors of Caxton.'[1]

Caxton evidently left his business and materials (perhaps having no son) to Wynkyn de Worde, who resumed printing as soon as possible, using Caxton's types 3, 4*, 6, 7, 8, his initials, his woodcuts and his device.[2] Duff blamed Wynkyn's lack of 'vigour or enterprise' for a fancied drop in production in 1492-3, while Plomer more chari-tably explained this by delay in the probate of Caxton's will. But in fact Wynkyn's 1493 was as productive as any of Caxton's later years, for it began with the completion of a third edition of the vast *Golden Legend* in 436 leaves (enough for four or more ordinary books) on 20 May, and included also, at least, *Chastising of God's Children* (complete with titlepage), the companion *Treatise of Love*,[3] and a new quarto edition of *Festial* (in an elegant Parisian type, Wynkyn's first new fount). *Golden Legend* must have taken well over half a year, and was itself preceded by the substantial *Life of St. Katherine* (ninety-six leaves) and a reprint of Caxton's brief *Book of Courtesy*. So Wynkyn must have recommenced full production no later than mid-1492; but it does look as though a break of about six months followed Caxton's death. Other printers were quick to seize their opportunity. Gerard Leeu at Antwerp produced reprints of Caxton's *Jason*, 2 June 1492, *Paris and Vienne*, 23 June 1492, and *Chronicles of England*, 1493,[4] not necessarily as an act of piracy, for these could well have been commissioned as stopgaps by Caxton's executors or

[1] *Life of St. Katherine of Siena* was Wynkyn's edition of 1492. *Birth of Our Lady* is an alternative title frequent in manuscripts for Lydgate's *Life of Our Lady* printed by Caxton in 1483. These books were thoughtfully chosen, as St. Margaret's was proud of possessing both a Lady Chapel and a chapel to St. Katherine.

[2] Both 3 (of which Caxton had passed his worn stock to the Saint Albans printer in 1486) and 4* were in recasts. Availability of new supplies of these ancient types as late as 1492, when Veldener had vanished from business or life in 1486, is hard to explain; it seems easier to believe that Caxton had acquired them from Veldener with his other later types in 1486, but neglected to use them before his death. Cf. p. 182, note 1, above.

[3] 'Translated out of French into English the year of Our Lord 1493 by a person that is unperfect in such work . . . a right well disposed person . . . and also caused the said book to be emprinted'. The date of translation shows that *Treatise of Love*, and no doubt also its twin *Chastising*, were orders obtained after Caxton's death by Wynkyn; but Wynkyn's other earliest works may well have been already on Caxton's existing order list.

[4] Leeu was stabbed and died in December 1492 in a quarrel with an employee, the typecutter Henrick Lettersnider. In the colophon of his *Chronicles* Leeu is saluted in words echoed by the commentator on Caxton's death cited above, as 'a man of great wisdom in all manner of cunning, which now is come from life unto the death which is great harm for many a poor man'.

by Wynkyn himself. More significantly still, it can hardly be by coincidence that Richard Pynson began to print in London towards the beginning of 1492. One of Pynson's first books in that year was a new edition of Caxton's *Canterbury Tales*, in which he referred to 'my worshipful master William Caxton'. It seems likely that Pynson himself had trained and worked under Caxton, and left to set up his own business immediately after Caxton's death, perhaps taking with him compositors laid off by Wynkyn at that time. There was plenty of room for both men, for the demand for English printed books which Caxton had created had begun to exceed the supply. Wynkyn and Pynson printed busily and prospered for four decades. Each produced just over a hundred pieces by 1500. Wynkyn's score totalled more than eight hundred when he died in 1535, and Pynson's more than five hundred by his death in 1530.

One man, at least, was far from content with the conduct of Caxton's executors. Gerard Crop, tailor, 'son in law late to one William Caxton late of Westminster bookprinter', claimed that Caxton 'lying in his deathbed' had bequeathed him '£80 in ready money', which Caxton's executor Sir Richard Ward priest first agreed to pay before witnesses, but then refused 'unless he were thereto compelled by the spiritual law'. One of Crop's witnesses was no less respectable a character than Robert Stowell, the Abbey master mason, with the rank of esquire, warden of the Fraternity of Our Lady at Westminster in 1491, and Caxton's next door neighbour but one near the Chapter House ever since 1476. The Archdeacon's court at Westminster found in Crop's favour, whereupon Ward threw him into Bread Street gaol and kept him there 'ever since midsummer hitherto nor will suffer his own wife to come at him nor relieve him', as Crop complained in his petition to Chancellor Morton for a writ of *corpus cum causa*.[1] On 17 October 1494 Caxton's daughter Elizabeth was herself obliged to petition Chancellor Morton for release from prison, for Crop had brought a chancery suit against her for 'divorce upon precontract' (that is, on the ground of a previous valid betrothal to another man), and secured her arrest through actions brought by third parties for an alleged debt of £52.[2] On 7 May 1496 Gerard and Elizabeth appeared in St. Stephen's Chapel Westminster before two of Henry's most distinguished councillors, Richard FitzJames and Richard Hatton, and were legally separated. Gerard was to receive '20 printed *Legends* at 13s 4d a *Legend*'—that is, to the purported value of only one sixth of his original claim, but really worth no more than a twelfth, for the *Sarum Legenda* was then selling at St. Margaret's for a mere 6s 8d—and each was bound over

---

[1] The context suggests that Crop's petition dates from the late summer of 1492. In any case it cannot be later than 1493, as Crop does not address Morton by the title of Cardinal, which he received on 20 September 1493.

[2] The sum is twice misprinted by Crotch as £3 ('iii li.' instead of 'lii li.'). The document specifies that 17 October is a Friday, from which the year date 1494 can be deduced.

in the sum of £100 not to 'vex, sue or trouble other for any matter or cause them concerning for matrimony betwixt them before had, and every of them to live sole from other, except that the said Gerard shall moe find the means to have the love and favour of the said Elizabeth'. It is unlikely that Gerard did so, or even tried. In March 1497 the Court of King's Bench ordered an enquiry into his goods under the recent statute against conspirators, on a plea brought on 31 August 1496 by Richard Redeknape mercer, whose relatives Emond and William had been Caxton's associates long ago in the 1450s and 1460s. Perhaps Gerard had become involved in the machinations of Perkin Warbeck. Poor Crop has been sternly handled (as 'unworthy to have been the son-in-law of so famous a father') by scholars who felt intuitively that a daughter of Caxton could do no wrong; but the documents suggest that the family deprived him of a legacy to which he was lawfully entitled, and that even in those benighted days a resolute woman could achieve liberation.

So the world went on after Caxton's death. In 1492, the first year he did not live to see, his patroness Queen Elizabeth Woodville died at Bermondsey Abbey; Henry made the invasion of France for which Caxton had supplied propaganda, in successful pursuit of peace with cash, like Edward in 1475; Columbus discovered Cuba and Haiti, not America; the impostor Perkin Warbeck was acknowledged by Duchess Margaret as her nephew, the true Duke of York resuscitated from the Tower, and rightful King Richard IV of England. Some thought Perkin was really her son by Henry of Bergen, Bishop of Cambrai, or a bastard of her brother Edward begotten in his Flanders exile of 1470; but more probably she hoped if all went well to replace the puppet Perkin with her beloved Clarence's son. Caxton would have congratulated himself more than ever for going Tudor in the year of Lambert Simnel. In England they called the termagant Margaret Duchess Juno, because she brought strife to the Olympus of royalty. Five centuries later the blood of Edward's Woodville daughter Elizabeth of York still runs in Elizabeth II, and the name of William Caxton remains illustrious.

This is how printing came to England five hundred years ago, and this, as far as we can know, was the life of the man who brought it.

Notes on the Illustrations

PLATES

I (Frontispiece)

An author, presumably Caxton, presenting his book, presumably *Recuyell of the Histories of Troy*, to Margaret Duchess of Burgundy in her audience chamber, from an engraving in a copy of *Recuyell* which once belonged to Queen Elizabeth Woodville, wife of Edward IV. It is here argued that this engraving was copied from an illumination, attributable to the Master of Mary of Burgundy, in the lost presentation manuscript given by Caxton to Margaret as patroness of his book, and so provides the only authentic surviving portrait of Caxton himself. Note the initials of Margaret and her husband Charles the Bold Duke of Burgundy on the canopy of the throne, with their over-optimistic motto *Bien en auiengne* ('may good come of it'), the lapdog held by one of the five wimpled ladies in waiting, the pet monkey, the courtiers in Burgundian winkle-picker shoes, the sideboard with ewers, towels and fingerbowls, and the gentleman usher with rod and plumed Burgundian bowler hat waiting in the doorway to ush Caxton out. From the copy of *Recuyell of the Histories of Troy* in the Huntington Library, San Marino, California (HM 62222).

IIa (p. 84)

Anthony Earl Rivers presenting the manuscript of his *Dicts of the Philosophers* to Edward IV in the royal Palace of Westminster at Christmas 1477. The full-length figures from left to right are the scribe Haywarde, Rivers in full plate armour with a surcoat bearing his arms and the Sun of York, Richard Duke of Gloucester (the future Richard III), Chancellor Thomas Rotherham, King Edward IV, the little Edward Prince of Wales (the future Edward V whom Richard murdered in the Tower), and Edward's Queen Elizabeth Woodville, Rivers' sister. Lambeth Palace Library MS. 265, leaf 1 verso.

IIb (p. 84)

Caxton's type 3: 135G., in his *Advertisement for the Sarum Ordinal*, 1479. 'If it plese ony man spirituel or temporel to bye ony pyes of two and thre comemoracio[n]s of salisburi vse enpryntid after the forme of this prese[n]t lettre whiche ben wel and truly correct, late hym come to westmonester in to the almonesrye at the reed pale and he shall haue them good chepe. Supplico stet cedula.' The line endings are irregular. From the copy in the Bodleian Library, Oxford.

III (p. 85)

Caxton's type 1: 120B., from *Recuyell of the Histories of Troy*, his first book and the first printed book in the English language, produced at Bruges in 1474-5. 'Therfore I haue practysed & lerned at my grete charge and dispense to ordeyne this said book in prynte after the maner & forme as ye may here see, and is not wreton with penne and ynke as other bokes ben . . .' The line endings are irregular. The initial T was added by pen by a later, seventeenth-century owner. British Library C.11.c.1, leaf 351 recto.

WILLIAM CAXTON
IV (p. 100)

Caxton's type 2: 135B., from the first dated book printed in England, *Dicts of the Philosophers*, 18 November 1477. The translator Anthony Earl Rivers, Edward IV's brother-in-law and Caxton's patron, describes his pleasant voyage to Compostella in July 1473: 'til I come in to the Spaynyssh see there lackyng syght of all londes, the wynde beyng good and the weder fayr, Thenne for a recreac(i)on & a passyng of tyme I had delyte & axed to rede som(m)e good historye . . .' The line endings are irregular. This first state of type 2 is easily distinguished by the use of separate types for t and h. British Library IB.55005, leaf 2 recto.

V (p. 101)

Caxton's type 7: 84G., from the 1489 *Indulgence* in aid of war against the Turks: 'IOhannes De Gigliis alias de liliis Ap[osto]licus Subdiacon[us] Et in Inclito Regno Anglie fructuu[m] & prouentuu[m] camere ap[osto]lice debito[rum] Collector, Et Perse[us] de Maluiciis decanus Eccl[esi]e Sancti michael[is] de leproseto Bononien[sis] . . .' The type is a reduced version of type 5, and was used by Caxton only in this *Indulgence*. This copy was issued to Henry Bost, Provost of Eton, on 24 April 1489. British Library IA.55126.

VI (p. 180)

Caxton's type 8: 114G., from *Ars moriendi*, 1491. 'Here begynneth a lytyll treatyse schortely compyled and called ars moriendi, that is to saye the craft for to deye for the helthe of mannes sowle.' Caxton obtained the type from Paris, where the same design was used by various printers. Only the first four lines are in type 8, the rest of this page being in type 6. From the copy in the Bodleian Library, Oxford, leaf A1 recto.

ILLUSTRATIONS IN TEXT
1 (p. 55)

Death among the printers, the earliest known illustration of a printing press, from *Danse Macabre*, printed by Mathias Huss, Lyons, 18 February 1499. The pressman is arrested by Death at the moment of pulling the press; his colleague meanwhile is beating the ink balls to replenish them for the next application; the compositor, with his case of type, composing stick, and two-page forme on the bench beside him, is setting from copy propped up before him; and in the adjoining bookshop the shopman is halted in the act of displaying a volume of his wares. British Library IB.41735, leaf b1 recto.

2 (p. 109)

Caxton's type 2*: 135B., from *Mirror of the World*, printed towards April-May 1481. The text and woodcut show that medieval man was well aware of the sphericity of the earth and of the theory of terrestrial gravitation: 'whether it were grete or lytyl, eche stone shold come in to myddle of therthe, wythout euer to be remeuid from thens, But yf it were drawen away by force, And they shold holden them one aboute another for to take place eueriche in the myddle of therthe . . .' This second state of the type is easily distinguished by th joined in a single type. British Library IB.55040, leaf d7 recto.

3 (p. 127)

Caxton's type 4: 95B., from *Knight of the Tower*, 31 January 1484, giving a dreadful warning to us all from the example of Perrot Lenard, 'whiche was sergeaunt of the saide chirche that same yere laye with a woman vnder an Awter, in whiche place this myracle befelle, They were ioyned to geder as a dogge is to a bytche, And in this manere they were fou[n]den & taken, & so ioyned & knytted to geder they were all the hole day, in so moche that they of the chirch & of the Cou[n]trey had leyser ynough to see & behold them . . .' This first state of the type is easily distinguished by A with hairline at right, and w with head-loop. British Library IB.55085, leaf d2 recto.

NOTES ON THE ILLUSTRATIONS

4 (p.133)

Caxton's type 4*: 100B., from his second edition of *Canterbury Tales*, towards August- September 1483. The Wife of Bath is shown riding side-saddle, with a roguish twinkle and a peekaboo hat:

> *'Bold was her face fayr and rede of hewe*
> *She was a worthy womman al hyr lyue*
> *Husbondys at the chyrche dore hadde she fyue*
> *Wythoute other companye in youthe . . .'*

This second state of the type is easily distinguished by A without hairline and w without headloop. British Library G.11586, leaf b5 verso.

5 (p. 159)

Caxton's device, from the unique *Legenda ad usum Sarum* in the British Library, the first of three books in which it is found in its second state, with two short breaks in the bottom frameline at right. The *Legenda* was printed for Caxton by Guillaume Maynyal in Paris, 14 August 1488, and the woodcut device was added on a blank page after the finished volumes reached Westminster. It shows Caxton's initials WC with his merchant's mark, apparently adapted to form a numerical pun on the years (14)47 (when he probably took the freedom of the Mercers' Company) and '74 (when he printed his first book the *Recuyell of the Histories of Troy*). The floral and foliar decoration is in the style of the wood-cutter at Gouda in Holland who supplied Caxton's later woodcut initials, and the block for the device was probably made by him to Caxton's order. British Library IB. 40010, leaf z5 verso.

6 (p. 171)

Caxton's type 5: 113G., from *Doctrinal of Sapience*, printed towards August-September 1489, with a Crucifixion woodcut previously used in *Speculum vitae Christi* in 1486. 'And when thermyte had herde all this, he was delyuered of all the euyll tamptacyons that he had. & glorified god & hys obscure iugementis which ben as a grete sowolowe without bottome . . .' The type is a reduced and simplified version of type 3. British Library IB.55129, leaf B1 recto.

7 (p. 175)

Caxton's type 6: 120B., from *Eneydos*, printed soon after 22 June 1490, showing the famous story of the mercer Sheffelde, who 'cam in to an hows and axed for mete, and specyally he axyd after eggys. And the goode wyf answerde, that she coude speke no frenshe . . .' The type is a reduced and simplified version of type 2. British Library IB.55135, leaf A2 verso.

8 (p. 185)

Caxton's *Fifteen Oes*, printed towards the summer of 1491, in type 6 with a fine Crucifixion cut and woodcut border pieces from his new *Horae* series. 'O Ihesu endles swetnes of louyng soules, O Ihesu gostly ioye passing & excedyng all gladnes and desires . . . *Fifteen Oes*, together with the lost sixth edition of *Horae ad usum Sarum* to which it belonged as a supplement, was probably Caxton's last book. British Library IA.55144, leaves 1 verso and 2 recto.

Select Bibliography

*Each entry is preceded by the abbreviated form
used for the references given in the Notes.*

Ames-Herbert   Ames, Joseph. *Typographical Antiquities.* 2nd edn. augmented by W. Herbert. 3 vols. 1785-90.

Anglo   Anglo, S. The British History in early Tudor propaganda, in *John Rylands Library Bulletin* xLIV (1961) 17-48.

Aurner   Aurner, Nellie S. *Caxton, mirrour of fifteenth-century letters.* 1926.

Barker   Barker, Nicolas. The Real Jane Shore, in *Etoniana* 4 June 1972.

Bean   Bean, J. M. W. *The Decline of English Feudalism.* 1968.

Beaven   Beaven, A. B. *The Aldermen of the City of London.* 2 vols. 1908.

Bennett, H. S.   Bennett, Henry S. *English Books and Readers 1475 to 1557.* 1969.

Bennett, J. A. W.   Bennett, J. A. W. Caxton and Gower, in *Modern Language Review* xLV (1950) 215-16.

Birch   Birch, J. G. William Caxton's stay at Cologne, in *The Library* ser. 4 IV (1923-4) 50-2.

Blades   Blades, William. *The Life and Typography of William Caxton.* 2 vols. 1861, 63.

Blades (1877)   —, —. *The Biography and Typography of William Caxton.* 1877; reprint ed. J. Moran, 1971.

Blades (1882)   —, —, —. 2nd edn. 1882.

Blake (World)   Blake, Norman F. *Caxton and his World.* 1969.

Blake (Prose)   —, —. *Caxton's own Prose.* 1973.

Blake (Sel)   —, —. *Selections from William Caxton.* 1973.

Blake (Chaucer)   —, —. Caxton and Chaucer, in *Leeds Studies in English* new ser. I (1967) 19-36.

Blake (Documents)   —, —. Two new Caxton Documents, in *Notes and Queries* ccxII (1967) 86-7.

Blake (Gower)   —,—. Caxton's Copy-text of Gower's *Confessio amantis*, in *Anglia* Lxxxv (1967) 282-93.

Blake (JR)   —, —. Investigations into the Prologues and Epilogues by William Caxton, in *John Rylands Library Bulletin* xLIx (1967) 17-46.

Blake (Knight)   —, —. The Noble Lady in Caxton's *The Book of the Knight of the Tower*, in *Notes and Queries* ccx (1965) 92-3.

Blake (Mercers)   —, —. Some Observations on William Caxton and the Mercers' Company, in *The Book Collector* xv (1966) 283-95.

Blake (Mirror)   —, —. The Mirror of the World and MS. Royal 19.A.IX, in *Notes and Queries* ccxII (1967) 205-7.

Blake (Reynard)   —, —, ed. *Caxton's History of Reynard the Fox.* 1970. EETS o.s. 263.

Blake (Suffolk)   —, —. William Caxton and Suffolk, in *Suffolk Institute of Archaeology Proceedings* xxIx (1962) 139-53; xxx (1964) 112-15.

BMC   *Catalogue of Books printed in the XVth Century now in the British Museum.* Photographic reprint. 1963, 67.

Bone   Bone, Gavin L. Extant Manuscripts printed from by Wynkyn de Worde, in *The Library* ser. 4 xII (1931-2) 284-306.

Bradshaw   Bradshaw, Henry. *Collected Papers.* 1889.

Brie (Brut)   Brie, F. W. D., ed. *The Brut or the Chronicles of England.* 2 vols. 1906, 08. EETS o.s. 131, 136.

Brie (Gesch) —, —. *Geschichte und Quellen der mittelenglischen Prosachronik, The Brute of England.* 1905.
Bühler (Book) Bühler, Curt F. *The Fifteenth-Century Book.* 1960.
Bühler (Dicts) —, —, ed. *The Dicts and Sayings of the Philosophers.* 1941. EETS o.s. 211.
Bühler (37) —, —. Caxton Variants, in *The Library* ser. 4 xvii (1937) 62-9.
Bühler (40) —, —. Two Caxton Problems, in *The Library* ser. 4 xx (1940) 226-71.
Byles Byles, Alfred T. P., ed. *The Book of the Fayttes of Armes and of Chyvalrye.* 1932. EETS o.s. 189.

Calendar (Close) *Calendar of Close Rolls*, Henry VI, iv-Henry VII, i. 1937-63.
Calendar (Milan) *Calendar of State Papers, Milan*, ed. A. B. Hinds. 1913.
Calendar (Papal) *Calendar of Papal Registers relating to Great Britain and Ireland*, xiii, xiv. 1955, 60.
Calendar (Patent) *Calendar of Patent Rolls*, Edward IV-Henry VII. 1897-1963.
Calendar (Venice) *Calendar of State Papers, Venice*. i. 1864.
Carpenter Carpenter, Edward. *A House of Kings. The history of Westminster Abbey.* 1966.
Carton Carton, C. Colard Mansion et les imprimeurs brugeois du xve siècle, in *Annales de la Société d'émulation pour l'étude de l'histoire et des antiquités de la Flandre* v (1847) 333-72.
Carus-Wilson Carus-Wilson, Eleanora M. *Medieval Merchant Venturers.* 1967.
Carus-Wilson-Coleman Carus-Wilson, E. M. and Coleman, Olive. *England's Export Trade 1257-1547.* 1963.
Childe The Childe of Bristowe, in *Camden Miscellany* iv (1859).
Chrimes Chrimes, S. B. *Henry VII.* 1972.
Clive Clive, Mary. *This Sun of York. A biography of Edward IV.* 1973.
CMH *Cambridge Medieval History.* viii. *The Close of the Middle Ages.* 1959.
CP *Complete Peerage of England*, ed. G. E. C(okayne). 1910-59.
Conway Conway, Agnes E. The Maidstone Section of Buckingham's Rebellion, in *Archaeologia Cantiana* xxxvii (1925) 97-119.
Cooke-Wordsworth Cooke, William, and Wordsworth, C., ed. *Ordinale Sarum sive Directorium sacerdotum.* 2 vols. 1901-2.
Cousland Cousland, C. W. The Hand of Caxton, in *Penrose Annual* xxxix (1937) 54-6.
Croft Croft, P. J. *Lady Margaret Beaufort, Elizabeth of York, and Wynkyn de Worde.* 1958.
Crosland Crosland, J. *Sir John Fastolfe.* 1970.
Crotch Crotch, Walter J. B. *The Prologues and Epilogues of William Caxton.* 1928. EETS o.s. 176.
Crotch (1930) —, —. William Caxton, an Englishman of the fifteenth century, in *Economica* x (1930) 56-73.
Culley-Furnivall Culley, W. T. and Furnivall, F. J. ed. *Caxton's Eneydos.* 1890. EETS e.s. 57.

Davies Davies, Katharine. *The First Queen Elizabeth (Elizabeth Woodville).* 1937.
De Ricci De Ricci, Seymour. *A Census of Caxtons.* 1909.
DNB *Dictionary of National Biography.*
Dodgson Dodgson, Campbell. English Devotional Woodcuts of the late fifteenth century, in *Seventeenth Volume of the Walpole Society* (1929) 95-108.
Doyle Doyle, A. I. The Work of a late fifteenth-century English scribe, William Ebesham, in *John Rylands Library Bulletin* xxxix (1956-7) 298-325.
Duff (XV) Duff, Edward G. *Fifteenth Century English Books. A bibliography.* 1917.
Duff (Century) —, —. *A Century of the English Book Trade 1457-1557.* 1905.

Duff (*Comm*) —, —, ed. *Commemoracio lamentacionis siue compassionis Beate Marie.* 1901.

Duff (*E Chan P*) —, —. Early Chancery Proceedings concerning members of the book trade, in *The Library* ser. 2 VIII (1907) 408-20.

Duff (*EEP*) —, —. *Early English Printing. A series of facsimiles.* 1896.

Duff (*E Prov*) —, —. *The English Provincial Printers to 1557.* 1912.

Duff (*Horae*) —, —. *Horae beatae virginis Marie secundum usum Sarum, the unique copy printed at Westminster by Caxton circa 1477.* 1908.

Duff (*Westminster*)—, —. *The Printers, Stationers and Bookbinders of Westminster and London from 1476 to 1535.* 1906.

Duff (*WC*) —, —. *William Caxton.* 1905.

Edwards Edwards, H. L. R. *Skelton.* 1949.

Ekwall Ekwall, E. *Concise Oxford Dictionary of English Place-names.* 1936.

Emden (*Camb*) Emden, A. B. *Biographical Register of the University of Cambridge to 1500.* 1963.

Emden (*Oxford*) —, —. *Biographical Register of the University of Oxford to 1500.* 3 vols. 1957-9.

Excerpta Hist. Excerpta Historica. ed. S. Bentley. 1831.

Fisher Fisher, John H. *John Gower.* 1965.

Förster Förster, M. Ueber Benedict Burghs Leben und Werke, in *Archiv für das Studium der neueren Sprachen* CI (1898) 29-64.

Francis Francis, Sir Frank C. *Robert Copland, sixteenth-century printer and translator.* 1961.

Gallagher Gallagher, J. E. The Sources of Caxton's *Ryal Book* and the *Doctrinal of Sapience,* in *Studies in Philology* LXII (1965) 40-62.

Girling Girling, F. A. *English Merchants' Marks.* 1964.

Goff Goff, Frederick R. *Incunabula in American Libraries. A third census.* 1964; reprint 1974.

Greg Greg, Sir Walter W. The Early Printed Editions of Canterbury Tales, in *Proceedings of the Modern Language Association* XXXIX (1924) 737-61.

Grierson Grierson, Philip. The Dates of the *Livre des mestiers* and its derivatives, in *Revue Belge de philologie et d'histoire* XXXV (1957) 778-83.

GW Gesamtkatalog der Wiegendrucke. 1925-40, 1972- .

Hare Hare, W. Loftus. A newly discovered volume printed by William Caxton, in *Apollo* XIV (October 1931) 205-13.

Hasted Hasted, Edward. *History and Topographical Survey of the County of Kent.* 4 vols. 1778-99.

Heilbronner Heilbronner, Walter L. *Printing and the Book in Fifteenth-Century England. A bibliographical survey.* 1967.

Hellinga (*Notes*) Hellinga, Lotte. Notes on the order of setting a fifteenth-century book, in *Quaerendo* IV (1974) 64-9.

Hellinga (*Ovid*) Hellinga, Lotte, and W. *Colard Mansion: an original leaf from the Ovide Moralisé,* Bruges, 1484. 1963.

Herrtage Herrtage, Sidney J. H., ed. *Charles the Grete.* 1881. EETS o.s. 59.

Hodnett Hodnett, Edward. *English Woodcuts 1480-1535.* 1935.

Hommel Hommel, Luc. *Marguerite d'York.* 1959.

HPT Hellinga, Wytze and Lotte. *The Fifteenth-Century Printing Types of the Netherlands.* 2 vols. 1966.

Imray Imray, Jean. The Merchant Adventurers and their Records, in *Journal of the Society of Archivists* II (1960-4) 457-67.

Jacob Jacob, Ernest F. *The Fifteenth Century 1399-1485.* 1961.

Jones Jones, Philip E. *Calendar of Pleas and Memoranda Rolls 1437-1457.* 1954.

Kendall. Kendall, Paul M. *Richard III.* 1955.
Kerling Kerling, Nellie J. M. Caxton and the trade in printed books, in *The Book Collector* IV (1955) 190-9.
Kingsford Kingsford, C. L. *English Historical Literature in the Fifteenth Century.* 1913.
Knocker Knocker, Herbert W. Letter in *The Times,* 24 April 1943 5d.
Knowles (C) Knowles, C. Caxton and his two French sources (in *Game of Chess*), in *Modern Language Review* XLIX (1954) 417-23.
Knowles (J) —, —. Jean de Vignay, in *Romania* LXXV (1954) 353-83.

Lander Lander, J. R. *The Wars of the Roses.* 1965.
Lander (M) —, —. Marriage and Politics in the Fifteenth Century, the Nevilles and the Wydevilles, in *Bulletin of the Institute of Historical Research* XXXVI (1963) 119-52.
Lathrop Lathrop, H. B. The Translations of John Tiptoft, in *Modern Language Review* XLI (1926) 496-501.
Lauritis Lauritis, J. A., ed. *A Critical Edition of John Lydgate's Life of Our Lady.* 1961.
Leach Leach, MacEdward, ed. *Paris and Vienne.* 1957. EETS o.s. 234.
Lenaghan Lenaghan, R. T., ed. *Caxton's Esope.* 1967.
Letts Letts, Malcolm. *Bruges and its Past.* 1924.
Letts (M) —, —. *Sir John Mandeville.* 1949.
Lyell-Watney Lyell, Laetitia, and Watney, F. D., ed. *Acts of Court of the Mercers' Company.* 1936.
Lysons. Lysons, D. *The Environs of London.* 4 vols. 1792-6.
Lyte Lyte, Sir H. C. Maxwell. *A History of Eton College.* 1899.

MacCracken MacCracken, Henry N., ed. *The Minor Poems of John Lydgate.* 2 vols. 1910, 34. EETS o.s. 107, 192.
McCusker McCusker, H. A Book from Caxton's Library, in *More Books* ser. 6 XV (1940) 275-84.
MacGibbon MacGibbon, David. *Elizabeth Woodville.* 1938.
Mackie Mackie, J. D. *The Earlier Tudors.* 1966.
Mancini Mancini, D. *The Usurpation of Richard the Third,* tr. & ed. C. A. J. Armstrong. 2nd edn. 1969.
Meier-Ewart Meier-Ewart, C. A Middle English Version of the *Fifteen Oes,* in *Modern Philology* LXVIII (1971) 355-61.
Michel Michel, H. *L'Imprimeur Colard Mansion et le Boccace de la Bibliothèque d'Amiens.* 1925.
Mitchell Mitchell, Rosamond J. *John Tiptoft.* 1938.
Mitchner Mitchner, Robert W. Wynkyn de Worde's use of the Plimpton Manuscript of *De proprietatibus rerum,* in *The Library* ser. 5 VI (1950-1) 194-6.
Morgan Morgan, Margery M. A Specimen of Early Printer's Copy, Rylands English MS 2, in *John Rylands Library Bulletin* XXXIII (1950-1) 194-6.
Morgan-Painter Morgan, Paul, and Painter, G. D. The Caxton *Legenda* at St. Mary's Warwick, in *The Library* ser. 5 XII (1957) 225-39.
Mortimer Mortimer, Jean E. An Unrecorded Caxton at Ripon Cathedral, in *The Library* ser. 5 VIII (1953) 37-42.
Mulders Mulders, J. A., ed. *The Cordyal by Anthony Woodville Earl Rivers.* 1962.
Myers Myers, A. R. *English Historical Documents 1327-1485.* 1969.
Myers (Household) —, —, ed. *The Black Book of the Household of Edward IV.* 1959.

Nichols Nichols, J. *Illustrations of manners and expenses in England.* 1797.
Nixon Nixon, Howard M. A Binding from the Caxton Bindery, in *The Book Collector* XIII (1965) 52.

Oates Oates, John C. T. *A Catalogue of the Fifteenth-Century Printed Books in the University Library, Cambridge.* 1954.

WILLIAM CAXTON

Oates-Harmer —, —, and Harmer, L., *ed. Vocabulary in French and English. A facsimile of Caxton's edition.* 1964.
Offord Offord, M. Y., *ed. The Book of the Knight of the Tower.* 1971. EETS suppl. ser. no. 2.

Pächt Pächt, Otto. *The Master of Mary of Burgundy.* 1948.
Painter (Caxton) Painter, G. D. Caxton through the Looking-Glass, in *Gutenberg Jahrbuch* 1963 73-80.
Painter (Gutenberg) —, —. Gutenberg and the B36 Group, in D. E. Rhodes, *ed. Essays in honour of Victor Scholderer.* 1970.
Paston The Paston Letters, ed. J. Gairdner. 6 vols. 1904.
Pays-Bas Le Cinquième Centenaire de l'imprimerie dans les Anciens Pays-Bas. *Exposition à la Bibliothèque royale Albert 1er. Catalogue.* 1973.
Pearce Pearce, E. H. *The Monks of Westminster.* 1916.
Peartree Peartree, S. M. A Portrait of William Caxton, in *Burlington Magazine* VII (1905) 383-7.
Pfaff Pfaff, R. W. *New Liturgical Feasts in later medieval England.* 1970.
Plomer (Caxton) Plomer, Henry R. *William Caxton.* 1925.
Plomer (Worde) —, —. *Wynkyn de Worde & his contemporaries.* 1925.
Pollard, A. F. Pollard, A. F. *The Reign of Henry VII from contemporary sources.* 3 vols. 1913, 14.
Pollard, A. W. (Ind) Pollard, Alfred W. The New Caxton Indulgence, in *The Library* ser. 4 IX (1928-9) 86-9.
Pollard, A. W. (Rec) —, —. Recent Caxtoniana, in *The Library* ser. 2 VI (1905) 337-53.
Pollard, G. Pollard, Graham. The Company of Stationers before 1557, in *The Library* ser. 4 XVIII (1937-8) 1-38.
Pollard-Ehrman Pollard, Graham, and Ehrman, Albert. *The Distribution of Books by Catalogue.* 1965.
Pollet Pollet, Maurice. *John Skelton.* 1971.
Power-Postan Power, Eileen, and Postan, M. M. *Studies in English Trade in the Fifteenth Century.* 1951.
Preston Preston, A. E. *The Church and Parish of St. Nicholas Abingdon.* 1929.

Reaney Reaney, P. H. *The Place-Names of Cambridgeshire.* 1943.
Ringler Ringler, William A. A Bibliography and First-line Index of English Verse printed through 1500, in *Papers of the Bibliographical Society of America* XLIX (1955) 153-80.
Ritson Ritson, J. *Ancient Songs.* 1790.
Ropp Ropp, G. von der. *Hanserecesse 1431-1476.* Abt. 2. 7 vols. 1876-92.
Rot. Parl. *Rotuli Parliamentorum.* 7 vols. 1767-1832.
Routh Routh, E. M. G. *Lady Margaret. A memoir of Margaret Beaufort.* 1924.
Ruysschaert (AFH) Ruysschaert, J. Lorenzo Guglielmo Traversagni de Savona, in *Archivum Franciscanum Historicum* XLVI (1955) 195-210.
Ruysschaert (JRL) —, —. Les Manuscrits autographes de deux oeuvres de Lorenzo Guglielmo Traversagni imprimées chez Caxton, in *John Rylands Library Bulletin* XXXVI (1953-4) 191-7.
Rymer Rymer, Thomas. *Foedera.* 20 vols. 1704-35.

Salter Salter, F. M. Skelton's *Speculum Principis*, in *Speculum* IX (1934) 25-37.
Schanz Schanz, Gustav. *Englische Handelspolitik gegen Ende des Mittelalters.* 2 vols. 1881.
Schirmer Schirmer, Walter F. *John Lydgate.* 1961.
Scholderer (Fifty) Scholderer, Victor. *Fifty Essays in Fifteenth- and Sixteenth-Century Bibliography.* ed. D. E. Rhodes. 1966.
Scholderer (Wales) —, —. *Handlist of Incunabula in the National Library of Wales.* 1940.
Scofield Scofield, Cora L. *The Life and Reign of Edward the Fourth.* 2 vols. 1923.

SELECT BIBLIOGRAPHY

Scott   Scott, Edward J. L. Letters in *The Athenaeum* 25 Dec. 1880 867; 10 June 1893 734; 13 April 1895 474; 25 March 1899 371; 10 Feb. 1900 177.
Sharpe   Sharpe, R. R. *Calendar of Letter Books A-L.* 1912.
Shaw   Shaw, Sally. Caxton and Malory, in J. A. W. Bennett, *ed. Essays on Malory.* 1963.
Sheppard   Sheppard, Leslie A. A new Light on Caxton and Colard Mansion, in *Signature* new ser. xv (1952) 28-39.
Simpson   Simpson, Percy. *Proof-reading in the sixteenth, seventeenth and eighteenth centuries, ed.* H. Carter. 1970.
Skeat   Skeat, T. C. The Caxton Deeds, in *British Museum Quarterly* xxviii (1964) 12-15.
Smit   Smit, Homme J. *Bronnen tot de geschiedenis van den Handel met Engeland, 1435-85.* 1928.
Sotheby   *Sotheby Sales Catalogues, various dates.*
STC   *Short-title Catalogue of Books printed in England 1475-1640.* By A. W. Pollard and G. R. Redgrave. 1926.
Stein   Stein, Walther. *Hansisches Urkundenbuch X 1471-85.* 1907.
Stevenson (Briquet)   Stevenson, Allan H. Introduction to C. M. Briquet, *Les Filigranes* i. 1968.
Stevenson (Caxton)   —, —. *Caxton and the Unicorns.* 1967.
Stevenson (Louvain)   —, —. The First Book printed at Louvain, in D. E. Rhodes, *ed., Essays in honour of Victor Scholderer.* 1970.
Stevenson (Tate)   —, —. Tudor Roses from John Tate, in *Studies in Bibliography* xx (1967) 15-34.
Stow   Stow, John. *A Survey of London. ed.* C. L. Kingsford. 2 vols. 1908, 27.
Stowe   *Catalogue of the Stowe Manuscripts in the British Museum.* 1895.

Tafur   Pero Tafur. *Travels and adventures 1435-1439. tr.* & *ed.* M. Letts. 1926.
Tanner   Tanner, Lawrence E. William Caxton's Houses at Westminster, in *The Library* ser. 5 xii (1957) 153-66.
Thielemans   Thielemans, Marie R. *Bourgogne et Angleterre. Relations politiques et économiques 1435-1467.* 1966.
Thomas   Thomas, Arthur H. *Calendar of Pleas and Memoranda Rolls* 1413-37. 1943.
Thomas (H)   Thomas, Sir Henry. *Wilh. Caxton uyss Engelant. Evidence that the first English printer learned his craft at Cologne.* 1928.
Thrupp   Thrupp, Sylvia L. *The Merchant Class of Medieval England* 1300-1500. 1949.

Unger   Unger, W. S. *Bronnen tot de geschiedenis van Middelburg.* 3 vols. 1923-31.

Van der Linden   Van der Linden, H. *Itinéraires de Charles duc de Bourgogne, Marguerite de York et Marie de Bourgogne.* 1936.
Van Praet (Mansion)   Van Praet, Joseph B. B. *Notice sur Colard Mansion.* 1829.
Van Praet (Gruthuyse)   —, —. *Recherches sur Louis de Bruges, seigneur de Gruthuyse.* 1831.
Vaughan (Charles)   Vaughan, Richard. *Charles the Bold.* 1973.
Vaughan (Philip)   —, —. *Philip the Good.* 1970.
VCH   *Victoria County History.*
Vickers   Vickers, Kenneth H. *England in the Later Middle Ages.* 1961.
Vinaver   Vinaver, Eugene, *ed. The Works of Sir Thomas Malory.* 2nd edn. 3 vols. 1967.

Walcott   Walcott, M. E. C. *Westminster.* 1851.
Watney   Watney, Sir J. *History of the Hospital of St. Thomas of Acon.* 1892.
Weale (Beffroi)   Weale, W. H. J. Documents inédits sur les enlumineurs de Bruges, in *Le Beffroi* iv (1872-3) 238-337.
Weale (EPB)   —, —. Early Printing at Bruges, in *Transactions of the Bibliographical Society* iv (1898).

WILLIAM CAXTON

Webb   Webb, Christopher A. Caxton's *Quattuor sermones*. A newly discovered edition, in D. E. Rhodes, *ed.*, *Essays in honour of Victor Scholderer*. 1970. 407-25.

Wedgwood   Wedgwood, Josiah C. *History of Parliament. Biographies 1439-1509*. 1936.

Weiss (Humanism)   Weiss, Roberto. *Humanism in England during the Fifteenth Century*. 1957.

Weiss (Tiptoft)   —, —. The Library of John Tiptoft, in *Bodleian Quarterly Record* VIII (1935-7) 157-64.

Weiss (Gigli)   —, —. Lineamenti di una biografia di Giovanni Gigli, 1434-1498, in *Rivista di storia della Chiesa in Italia* I (1947) 379-91.

Westlake (St. M)   Westlake, Herbert F. *St. Margaret's Westminster*. 1914.

Westlake (WA)   —, —. *Westminster Abbey*. 2 vols. 1923.

Williams, C. H.   Williams, C. H. The Rebellion of Humphrey Stafford, in *English Historical Review* XLIII (1928) 181-9.

Williams, N.   *The Life and Times of Henry VII*. 1973.

Wilson (Chess)   Wilson, R. H. Caxton's Chess Book, in *Modern Language Notes* LXII (1947) 93-102.

Wilson (Esope)   —, —. The Poggiana in Caxton's Esope, in *Philological Quarterly* XXX (1951) 348-52.

Wordsworth   Wordsworth, Christopher, *ed. The Tracts of Clement Maydeston, with the remains of Caxton's Ordinale*. 1894.

Wordsworth-Littlehales   Wordsworth, Christopher, and Littlehales, H. *Old Service Books of the English Church*. 2nd edn. 1910.

Workman   Workman, S. K. Versions by Skelton, Caxton and Berners of a prologue by Diodorus Siculus, in *Modern Language Notes* LVI (1941) 252-8.

References

The references for each paragraph are grouped separately, preceded by the page number and first words of the paragraph. For the full titles of the books or articles cited, see Select Bibliography.

1 THE FAMILY

Page and paragraph

1 *Every* . . . *Crotch* 4, 96.

2 *There is* . . . *Crotch* xxix-xxx, cv; *Ekwall*; *Reaney* 157.

2 *Cawston* . . . *Crotch* xxx, xxxii; *Ekwall.*

3 *This William* . . . *Blades* I 2, 87-8, 269, 271-4; *Blades* (1877) 145; *Blake* (*World*) 14; *Crotch* xxxi-xxxii; *Jones* 172, 203; *Scott* 1880, 1893; *Sharpe* 176, 305; *Thomas* 276.

4 *Meanwhile* . . . *Blades* I 3, 270-5; *Hasted* II 317; *Plomer* (*Caxton*) 15-18; *VCH Kent* III 403-5.

4 *Perhaps* . . . *Blades* I 90, 270; *Crotch* xxxiii-xxxv; *Knocker*; *Lyell-Watney* 49, 58-9, 109, 140; *Scott* 1895; *Sotheby* 23 Feb. 1959 no. 208a.

5 *The family* . . . *Blades* I 119-22, 270, 274; *Crotch* xxxi-xxxiv, cv, cvii, cxvii, cl; *Jones* 179; *Pearce* 165; *Scott* 1893, 1895; *Tanner* 153; *Thomas* 203.

6 *Our tally* . . . *Blades* I 269, 273-4; *Blades* (1877) 3; *Blake* (*Suffolk*); *Duff* (*WC*) 15; *Skeat*; *Thomas* 300.

2 THE APPRENTICE

8 *The first* . . . *Blades* I 3-4, 88, 101, 105-8; *Blake* (*Mercers*) 285; *Duff* (*WC*) 16; *Plomer* (*Caxton*) 20.

8 *It was* . . . *Blades* I 2, 4, 88; *Blades* (1877) 6; *Blake* (*World*) 22-4; *Blake* (*Mercers*); *Crotch* xxvii-xxviii, xliv; *Duff* (*WC*) 16; *Plomer* (*Caxton*) 18-20.

9 *Blake's* . . . *Blades* I 86-8, 101-2; *Blake* (*World*) 23.

10 *All* . . . *Crotch* xxviii.

10 *The Mercers'* . . . *Blake* (*World*) 27; *Carus-Wilson* xxi-xxx, 151; *Imray*; *Lyell-Watney* viii-ix, xvi.

11 *The alderman* . . . *Blades* I 5, 87-8.

11 *Large's house* . . . *Blades* I 5-6; *Lyell-Watney* viii; *Stow* I 278; *Watney* 5-6.

11 *Large's household* . . . *Blades* I 3-4, 6-7, 87-8, 99-104; *Crotch* xl.

12 *So* . . . *Blades* I 5, 85-6.

12 *During* . . . *Blades* I 231, 241-4; *Kent* 540.

13 *The witch* . . . *Blades* I 237, 243-4.

13 *Robert* . . . *Blades* I 6-7, 95-104; *Blades* (1877) 151-5.

14 *Not all* . . . *Blades* I 87, 105; *Stow* I 186.

14 *It is* . . . *Blades* I 87; *Carus-Wilson* 81-2; *Childe* 11, 27; *Crotch* xxxix-xl, clix; *Lyell-Watney* xi-xii.

15 *Such* . . . *Blake* (*World*) 18-19, 29-30; *Crotch* xxxviii, clix, 4.

3 THE MERCHANT

16 *Caxton* . . . *Blades* I 239, 241, 247, 255, 263; *Vaughan* (*Philip*) 75-84, 108-10.

17 *But even* . . . *Blades* I 249; *Jacob* 359; *Myers* 259; *Power-Postan* 27-8, 127-9; *Thielemans* 154-5, 337; *Vaughan* (*Philip*) 110.

17 *Our first* ... *Blades* I 11, 105-8; *Crotch* xliii-xliv.
18 *In the* ... *Beaven* II 229; *Blades* I 12, 88-9; *Blake* (*Mercers*) 292-5; *Blake* (*World*) 33, 65 (pl. 2); *Crotch* xliv-xlv, 99-100; *Lyell-Watney* x; *Stow* II 172.
19 *The next* ... *Blades* I 89, 253; *Blake* (*Mercers*) 294-5; *Blake* (*World*) 65 (pl. 2), 219.
19 *At an* ... *Blake* (*World*) 38; *Carus-Wilson* xxxi, 170; *Power-Postan* 55; *Thielemans* 262.
20 *Caxton* ... *Barker* 389; *Blake* (*World*) 39, 84; *Calendar* (*Close*) 2 Henry VII no. 203; *Crotch* xlv-xlvi, cxxxiii-cxxxiv; *Plomer* (*Caxton*) 34.
20 *What* ... *Blake Documents*.
21 *Next spring* ... *Blake* (*Documents*); *Blake* (*World*) 39; *Crotch* ci-cii, cxlviii-cl.
21 *The documents* ... *Jacob* 362; *Lyell-Watney* xv; *Power-Postan* 66-70; *Thielemans* 250-1, 266-9.
22 *Bruges* ... *Blades* I 40, 58, 127; *Crotch* xlii; *Letts* 20-1, 27-32, 104-5, 112; *Tafur* 198, 200; *Vaughan* (*Philip*) 87-91, 136.
23 *Every* ... *Crotch* lxxxi; *Crotch* (1930) 60-1; *Letts* 10, 102, 109-10, 112, 139; *Tafur* 199-200.
23 *The English* ... *Carus-Wilson* xvii-xviii, 146, 153, 176; *Crotch* xlviii; *Letts* 10; *Lyell-Watney* xiii-xv; *Power-Postan* 20-30, 34-8, 150-3, 380; *Schanz* II 575.

4 THE GOVERNOR

25 *This* ... *Blades* I 262; *Carus-Wilson* xxii; *Crotch* (1930) 63; *Power-Postan* 2, 21, 24, 28-30, 121, 132, 376; *Scofield* I 9, 12, 89, 92, 121-2; *Smit* II 955, no. 1502.
25 *The governor* ... *Blake* (*World*) 20-1, 38; *Calendar* (*Milan*) I 43-4, 47, 106-8; *Crotch* xxxviii, xlviii-liii, lxxviii, clviii-clix; *Schanz* II 577, no. 24, 578, no. 30; *Scofield* I 19, 113, 130-2, 191, 233; *Smit* I 933-4; *Thielemans* 277-82, 370-1, 374-5, 493.
27 *Still* ... *Blades* I 13, 110-19.
28 *But Overey's* ... *Blades* I 15, 89; *Carus-Wilson* 153; *Crotch* xlix-liii, lxxii, lxxviii-lxxix; *Lyell-Watney* xiv, 297-8, 629; *Schanz* II 575, no. 5, 578, nos. 34, 35; *Smit* II 967 no. 1519; *Thielemans* 278, 389, 398.
28 *Overey's* ... *Crotch* (1930) 64-5; *Smit* II 970, no. 1524.
29 *Perhaps* ... *Blades* I 14-15; *Crotch* liv-lv; *Thielemans* 271.
29 *Perhaps Caxton* ... *Crotch* lvi-lvii, cxxxiv-cxxxv; *Smit* II 969, no. 1523; *Unger* I 317.
29 *A new* ... *Blades* I 90; *Scofield* I 31, 65, 147, 159-60, 169, 178, 210, 232, 265, 273, 292-3, 301-4, 312-13, 333-4, 384-5, 421, II 404-17; *Thielemans* 262, 379, 391.
30 *Caxton's duties* ... *Blades* I 89-90; *Crotch* lvi; *Lyell-Watney* 59; *Myers* 1042; *Scofield* I 229, 285-6.
31 *Another* ... *Scofield* I 211-12, 259, 306; *Thielemans* 374-5, 384-6, 389, 398-402.

5 THE DIPLOMAT

32 *The Anglo-Burgundian* ... *Jacob* 535-7; *Scofield* I 50-2, 355-8, 397-8.
32 *Amid* ... *Blades* I 124-5; *Myers* 1042-3, 1047-9; *Scofield* I 346, 356-8; *Thielemans* 413-14; *Vaughan* (*Philip*) 374-7.
33 *Officially* ... *Crotch* lviii-lix; *Smit* II 985, no. 1545, 1007, no. 1563; *Thielemans* 281; *Vaughan* (*Philip*) 132-5, 224-9.
33 *Thanks* ... *Blades* I 90-1; *Crotch* lix; *Lyell-Watney* 277; *Myers* 1100; *Power-Postan* 28, 346, 377, 408; *Thielemans* 208.
34 *Anglo-Burgundian* ... *Blades* I 91; *Crotch* lix-lx; *Lyell-Watney* 279-80; *Myers* 1100-1; *Scofield* I 367, 372, 378-80, 385-6, II 407; *Thielemans* 415-18; *Vaughan* (*Philip*) 391-7.

REFERENCES

34 *No solution* . . . *Scofield* I 404-6, 409; *Thielemans* 274, 419-24.
35 *Next year* . . . *Duff (Westminster)* 4; *Emden (Oxford)* III 1610; *Oates* 28;
   Pollard-Ehrman 23; *Scofield* I 404, 409, 412; *Thielemans* 418, 420;
   Vaughan (Charles) 45; *Weale* (EPB) 211.
35 *Meanwhile* . . . *Scofield* I 406, 411-13, 427, 429; *Vaughan (Charles)* 44.
36 *At this* . . . *Scofield* I 416, 429-33, 442, 446-7, 452, II 472.
36 *Margaret* . . . *Lander* 154-5, 159-61; *Myers* 1175; *Scofield* I 456-64, 481,
   550, 562; *Vaughan (Charles)* 61, 158-9, 239.
37 *The Bruges* . . . *Beaven* I 168, II 15; *Blades* I 89, 93, 108-10; *Blake (World)*
   36, 90, 159; *Clive* 240; *Crotch* xlv, lxxii-lxxiii, cxiii, cxxxviii, 113;
   Lyell-Watney xv, 61-2, 485-6; *Lysons* IV 75; *Scofield* I 446, 453, 455,
   462, 464, 474, 485-6; *Unger* II 353.
38 *Only the* . . . *Crotch* lxv-lxxiv, cxxxviii; *Jacob* 356-60; *Power-Postan* 132-4;
   Scofield I 196-7, 271-2, 327-8, 390-2, 465-9, 486-7; *Thielemans* 209.
39 *During* . . . *Crotch* 2, 4.
39 *It was* . . . *Scofield* I 486-8, 498, 501.
40 *Oddly* . . . *Power-Postan* 28, 346; *Thielemans* 484.
40 *Warwick's* . . . *Blades* I 126-7; *Scofield* I 484-5, 504, 507, 513-14, 529-34,
   546-8; *Vaughan (Charles)* 60.
41 *Edward's* . . . *The Genealogist* new ser. III (1886) 65; *Lander* 181; *Scofield* I
   550-1, 556, 562-3, 566-8; *Vaughan (Charles)* 66-72.

6 THE EXILE

43 *Caxton is* . . . *Crotch* lxxvii, cxxxviii; *Scofield* II 30; *Unger* II 355.
43 *During* . . . *DNB*: Cooke, Sir Thomas; *Jacob* 565; *Myers* 1047-9; *Power-
   Postan* 346; *Scofield* I 556-64.
44 *Caxton's* . . . *Blake (World)* 219; *Power-Postan* 45; *Rymer* XI 735; *Scofield* I
   430, 455, II 122, 295, 420.
45 *So* . . . *Crotch* 2-5.
45 *When* . . . *Blades* I 20; *Crotch* lxxx, 2-6; *Scofield* I 567; *Van der Linden* 29;
   Vaughan (Charles) 159.
46 *Meanwhile* . . . *Archaeologia Cantiana* VII (1868) 269; *Conway* 98; *Lander*
   180-1, 193-5; *Lyell-Watney* 138; *Myers* 1099; *Scofield* I 435, 576,
   586-95, II 1-2, 12-13, 20-3.
47 *Duke* . . . *Scofield* II 7-10, 15-16.
48 *But* . . . *Birch*; *Crotch* cxxxix; *Stein* 46, no. 74, n. 3; *Thomas* (H).
48 *Duchess* . . . *Crotch* lxxxvi; *Jacob* 359-60; *Power-Postan* 134-6, 378, 404-5,
   408; *Ropp* II:6 435, 475; *Scofield* I 487, 566, II 29.
49 *Such* . . . *Crotch* lxxxv-lxci, cxxxix-cxl; *Ropp* II:6 544; *Rymer* XI 739;
   Scofield II 31.

7 LEARNING TO PRINT AT COLOGNE

51 *Caxton's* . . . *Crotch* 6-7; *Scofield* II 14-15.
51 *Caxton finished* . . . *Crotch* lxxxiv, 2; *Duff (WC)* 25; *Tafur* 137; *Van der
   Linden*; *Vaughan (Philip)* 143-5, 271-2, 358-72.
53 *The demand* . . . *Crotch* 7.
53 *Caxton's* . . . *Crotch* lxxxiv.
54 *No doubt* . . . *BMC* IX 1.
54 *Three* . . . *BMC* I 178-213, 232-7, IX 1; *Crotch* lxxxvii; *Duff (WC)* 22.
56 *Modern* . . . *BMC* IX liii; *HPT* I 18.
56 *An alternative* . . . *BMC* I 212-13; *BMC* VIII lxxvii, 372; *BMC* IX liii;
   HPT I 18-19; *Stevenson (Briquet)* *29.
57 *The Bartholomaeus* . . . *BMC* I 191, 216-17, 232-6, 240.

8 PRINTING AT BRUGES I: MARGARET AND CLARENCE

59 *Caxton's last* . . . *Crotch* cxxxix; *Thomas* (H).
59 *Caxton's next* . . . *BMC* IX 130; *Crotch* 16; *HPT* II 395; *Stevenson (Caxton)*.

60 *The natural* ... *BMC* ix liii; *HPT* i 25-8, 63-7; *Stevenson* (*Briquet*) *29-31; *Stevenson* (*Louvain*) 402-6.

61 *On the* ... *Blades* ii pl. xi, xii; *Duff* (*Century*) 24; *Duff* (*Westminster*) 5; *HPT* i 19, 21, 23-4, 64, 66.

62 *So, towards* ... *Blades* i 29-30; *BMC* i 8, 39; *BMC* iv 19; *BMC* viii xxi; *Crotch* 7-8; *De Ricci* no. 3.

63 *Caxton's first* ... *Blades* ii 6; *BMC* ix 130; *De Ricci* 3:11; *Pächt* pl. v, 62; *Pollard, A. W.* (*Rec*).

64 *Caxton produced* ... *Blades* i 278-9, ii 9-10; *BMC* ix 130; *Knowles* (*C*); *Wilson* (*Chess*).

65 *Ever since* ... *Blades* i 17, 70; *Lander* 202; *Scofield* ii 5-7, 26-7, 29, 39, 58-60, 85-6, 188-90.

65 *Around* ... *Blades* ii 10; *Crotch* c, 5, 10-16; *Scofield* ii 208.

66 *Duchess* ... *Blake* (*Prose*) 86-7; *Crotch* c, 5, 14; *Van der Linden*.

67 *In fact* ... *Crotch* xci-xcv, cxl-cxlvii; *Myers* 1045-7; *Power-Postan* 135-7; *Scofield* ii 29-33, 46-7, 50-1, 63-84, 107-8.

68 *Meanwhile* ... *Crotch* cxl-cxlii; *Jacob* 572-6; *Scofield* ii 24, 107-8, 121, 148.

68 *This was* ... *Crotch* xcv-xcvi, cxliv-cxlv, 12, 16; *Scofield* ii 130-2; *Smit* ii no. 1767.

69 *Ordinarily* ... *Scofield* ii 117, 132-47.

70 *So, in* ... *Crotch* 1-11, 91; *HPT* i 22-3.

71 *Another* ... *Crotch* ci-cii, cxlviii-cl; *Duff* (*E Chan P*) no. 1; *Scofield* ii 94.

9 PRINTING AT BRUGES II: COLARD MANSION

72 *Immediately* ... *Blake* (*World*) 61.

72 *Colard* ... *Blades* i 40, 127-8; *BMC* ix li-lii, 132-4; *Crotch* xcvii-xcviii; *Michel* 12, 51-2; *Pays-Bas* 212-38; *Sheppard* 35; *Van Praet* (*Mansion*) 2-3, 70-82; *Weale* (*Beffroi*) 238-87, 307-8.

73 *It is* ... *Blades* i 54-5, 60-1; *HPT* i 23, 29; *Sheppard*.

73 *But the* ... *Pays-Bas* 212-16, 222-3, 226-7, 231, pl. 42, 48, 49; *Van Praet* (*Mansion*) 65.

75 *The technical* ... *Blades* i 42; *BMC* ix 31, 108, 129-33; *Painter* (*Gutenberg*) 295, 316; *Sheppard* 38-9.

76 *A unique* ... *Blades* i 43-4, 54-5; *Morgan-Painter* 231; *Scholderer* (*Fifty*) 269.

76 *Caxton's* ... *Blades* i 51, 58; *Crotch* cxxx; *Pays-Bas* 216, pl. 42.

77 *This strong* ... *Blades* i 40; *Carton* 371; *Michel* 11, 13-14; *Pays-Bas* 212; *Weale* (*Beffroi*) 261, 266.

77 *As we* ... *BMC* viii xxi, 234; *BMC* ix xlvi; *Stevenson* (*Briquet*) *27; *Stevenson* (*Caxton*).

78 *Cardinal* ... *Pays-Bas* 217-18.

79 *Last* ... *Blades* i pl. viii, ii pl. xiii; *Mulders* xiv, xxvii.

79 *Type 2* ... *BMC* ix 129, 135; *HPT* i 19, 21, 24, 47, 64, 66.

80 *Caxton intended* ... *BMC* ix 161-2; *HPT* i 24, 64.

80 *When Caxton* ... *Blades* i 58; *BMC* ix 131; *Doyle* 318; *Pays-Bas* 226; *Van Praet* (*Mansion*) 37.

81 *Caxton's Bruges* ... *Clive* 229-32; *HPT* i 17, 21, 29, 33, 34, 52; *Scofield* ii 156, 164-5, 171-2.

10 A SHOP IN THE ABBEY

82 *On 30 September* ... *Blake* (*World*) 82-3; *Carpenter* 79, 88-9; *Crotch* ciii, clii-cliii; *Tanner* 153-60; *Westlake* (*WA*) i 149, 151.

83 *But several* ... *Blades* i 122; *Crotch* cv-cvii; *Duff* (*Century*) xi-xii, 126; *Plomer* (*Worde*) 121-3; *Walcot* 140; *Westlake* (*St M*) 145, 149, 150.

83 *Perhaps* ... *Crotch* ciii-civ, cxlvii-cxlviii; *Scofield* ii 46, 50-1, 63-80, 116, 119, 171-2.

REFERENCES

83 *Caxton's first* . . . Pollard A. W. (*Ind*).
84 *The Pope's* . . . Calendar (*Papal*) XIII 50; Carpenter 90; Clive 250; Crotch 38; Emden (*Oxford*): Gigli, Sant; Pearce 148; Myers 695; Preston 80, 198, 289; Scofield II 99, 165, 167-8, 388-9.
84 *Perhaps* . . . Aurner 174; Duff (*Century*) 31-2; Francis; Stevenson (*Briquet*) *27; Stevenson (*Caxton*).
85 *Caxton's first* . . . Crotch 33-4; Vaughan (*Philip*) 57, 137-9, 144, 152, 157, 160-2, 340.
86 *As Edward* . . . Carpenter 89; Crotch 34; Scofield I 546, II 54-6.
86 *The patron* . . . Clive 243-4; Crotch 111-12; Scofield I 260, 419, 521, II 3-4, 31-4, 59, 94, 117, 123, 165, 203-6.
87 *This* . . . Bühler (*Dicts*) x-xi.
87 *After* . . . Blades II 37, 39; Bühler (*Dicts*) liii-lvi.
88 *Caxton takes* . . . Crotch 18-31.
88 *Caxton's elaborate* . . . Clive 258, 268; DNB: Rivers; CP: Rivers; Lander 218; Scofield II 174, 182-6, 197, 251, 347, 353.
89 *Towards* . . . Blades II pl. XXVI; Blake (*Prose*) 77; Blake (*World*) 221-2; Crotch cvii.
90 *Clarence* . . . Clive 232-9, 242-4; Lander 217-20; Scofield II 184-212.
90 *On 20* . . . Aurner 92; Crotch 32.
91 *Much* . . . Blake (*Chaucer*); Crotch 90-1; Greg.
91 *Perhaps* . . . Blake (*Chaucer*); Blake (*World*) 101-6; Crotch 94.
92 *Another* . . . Blake (*Chaucer*); Blake (*World*) 198-9; Crotch 36-7, 99-100; Emden (*Cambridge*): Surigonus; HPT II pl. 61; Weiss 139.
93 *Meanwhile* . . . Blades II 51, 54; CP: Essex; Crotch cxxv; Emden (*Oxford*): Burgh; Förster; MacCracken I xxviii; Plomer (*Caxton*) 160; Scofield II 419-28; Skeat; Tanner 160-1.
94 *Two more* . . . Blades I 45, 54, II 29; Doyle 308-10; HPT I 24; Scofield II 7, 239, 251, 284.
95 *In February* . . . Blake (*Prose*) 154; Clive 231, 258; Crotch 38-9; Scofield II 54-5, 165, 250; Stow II 121, 379.
96 *The Nova* . . . Hellinga (*Notes*); Mortimer; Ruysschaert (*AFH*); Ruysschaert (*JRL*).

11 AT THE RED PALE

98 *It is* . . . Blades I 119; Crotch cxi-cxii; Duff (*Century*) 173; Nichols 4; Scofield II 235, 241, 249-50; Scott 1899.
99 *The Ordinal* . . . Blades II 70; Wordsworth.
99 *The Advertisement* . . . Crotch cv-cvi, cxvi-cxvii, cl-clii, clvii; Tanner 161-3.
99 *A closer* . . . Crotch cvi, cxvi-cxvii, clvii; Pearce 147-8, 154-5; Tanner 161-3; Westlake I 146, 149, 151, 157.
100 *Once again* . . . Crotch cl-clii; Tanner 161-4.
101 *In 1480* . . . Duff (*WC*) 47-8; Duff (*Westminster*) 11-12.
101 *Caxton's new* . . . Blades II pl. XVIII.
101 *Caxton's decision* . . . Crotch 71; HPT I 54; Pays-Bas 227-38.
102 *Two* . . . Doyle 309-10; Duff (*WC*) 49; Pfaff 47, 50; Wordsworth-Littlehales 193.
103 *Doctrine* . . . Duff (*WC*) 49; Grierson; Oates-Harmer.
104 *Meanwhile* . . . Mortimer; Ruysschaert (*JRL*).
104 *Abbot* . . . Blades I 120, II 80; Calendar (*Papal*) XIII 80, 197; Duff (*XV*) nos. 204-10; Pollard, A. F. I 100-1; Pollard, A. W. (*Ind*); Scofield II 254, 265, 388, 440.
105 *Meanwhile* . . . Blades II 110; Brie (*Brut*) 524; Duff (*WC*) 50; Kingsford 99-100, 113-39; Scofield II 389-90.
106 *Caxton printed* . . . Blake (*Prose*) 69; Conway 98; Scofield I 435, II 283-97; Sotheby 27 June 1966 no. 318.
106 *On 18* . . . Crotch 40; De Ricci 29, 30, 35.

12 FRIENDS IN COURT

108 *Caxton translated* . . . *Crotch* 50-9; *Scofield* I 276-9, 294, 296, 314-16.
108 *Caxton* . . . *Blades* II 85; *Blake* (*Mirror*); *Blake* (*World*) 35-6; *Hodnett* 1-2.
110 *Not all* . . . *Blake* (*World*) 127; *Crotch* 56.
110 *Mirror* . . . *Blake* (*Prose*) 12-18.
111 *Caxton's* . . . *Blake* (*Reynard*) xx, xlviii; *Crotch* 60-2; *Pays-Bas* 289.
111 *Immediately* . . . *Blades* II 91; *BMC* IX 130, 132-3; *Bühler* (*Dicts*) xxxix-xlvi; *Crotch* 41-4; *DNB*: Fastolf; *Doyle* 298, 307; *HPT* I 29; *Myers* 1205-6.
112 *As for* . . . *BMC* IX 133; *Bühler* (*Dicts*) xlvi; *Crotch* 44-7; *Van Praet* (*Mansion*) 54; *Weiss* 112-19.
113 *However* . . . *Crotch* 44-7, *Scofield* I 519-22, 546-7; *Weiss* 112-19.
113 *Who tipped* . . . *Clive* 79-81, 122, 127, 152, 157, 196; *Crotch* 43-4; *Excerpta Hist.* 241; *Kendall* 78; *MacGibbon* 87-8; *Mitchell* 120; *Scofield* I 232-3, 414-15, 437-9, 517, II 316.
114 *Caxton had* . . . *Blake* (*Prose*) 137-42.
115 *Caxton's propaganda* . . . *Blades* II 119; *Clive* 196, 263; *De Ricci* 46:2; *Scofield* II 3, 315, 320, 322, 389-90.
116 *In August* . . . *Clive* 258, 266, 273; *Scofield* I 176, II 332, 344-9, 353.
116 *Caxton's long* . . . *Blades* II 128; *De Ricci* 49:81; *Duff* (*Century*) 1, 45, 127, 174; *Plomer* (*Worde*) 138, 183-6.
117 *Polycronicon* . . . *Blake* (*Prose*) 128-33; *Blake* (*World*) 116; *BMC* IX 136; *Crotch* 64-9; *DNB*: Tinmouth.
117 *Just as* . . . *Crotch* 66-7.
118 *No doubt* . . . *Crotch* 71.
118 *Three* . . . *Blades* II 81-2; *Crotch* 11-12; *Hodnett* 1-2.
119 *Curia* . . . *Clive* xvi; *MacCracken* I xxxv; *Schirmer* 277.
120 *Caxton's fourth* . . . *Stevenson* (*Tate*) 25.
120 *Caxton's other* . . . *Duff* (*Horae*).

13 UNDER WHICH KING?

121 *Edward* . . . *Clive* 245-6, 277-9; *Lander* 224, 227; *Scofield* II 356, 364-8.
121 *To Caxton* . . . *Kendall* 183, 207-8; *Lander* 300; *Ritson* 87-8.
122 *Gloucester* . . . *Clive* 158, 288-9; *Kendall* 221; *Lander* 243-4; *Mancini* 133; *Scofield* II 161, 213.
122 *So Caxton* . . . *Lander* 241.
122 *But one* . . . *Clive* 285; *Kendall* 178-9; *Lander* 235-6; *Mancini* 117, 120.
123 *In the* . . . *Ames-Herbert* I 42; *Blades* II 131; *Catalogus Bibliothecae Harleianae* III 126 no. 1565; *De Ricci* 75:3; *Hare*; *MacCracken* I xliii; *Schirmer* 121-2.
124 *Pilgrimage* . . . *HPT* I 48-9.
125 *John Mirk* . . . *Aurner* 128; *Blades* II 135; *DNB*: Mirk.
125 *As a* . . . *Webb*.
126 *On 2* . . . *Bennett, J. A. W.*; *Blake* (*Gower*); *Blake* (*World*) 131.
126 *His next* . . . *Offord* xvii-xviii, xxxiv-xliii, 12, 31-2, 59-60, 87-90.
128 *Caxton had* . . . *Blake* (*Knight*); *Chrimes* 22-9; *Kendall* 179, 200, 260-1, 265-6, 268, 272-3, 286-7; *Lander* 250-1.
131 *Perhaps Curial* . . . *Blades* II 168; *Blake* (*World*) 93; *Crotch* 39, 89.
131 *Likewise* . . . *Crotch* 69.
132 *Another* . . . *Crotch* 85; *De Ricci*, 71, 72; *Duff* (*WC*) 57; *Duff* (*XV*) 266, 266a.
132 *Canterbury* . . . *Blades* (1882) 293; *Crotch* 90-1; *Hodnett* 3.
134 *It is* . . . *Blake* (*Chaucer*); *Blake* (*World*) 104; *British Museum Quarterly* xxv (1963-4) 100-1; *Greg*.
135 *To the* . . . *DNB*: Carmelianus; *Weiss* 169-72.
135 *Did Carmelianus* . . . *Blades* (1882) 268; *CMH* VIII 196-7; *Duff* (xv) pp. 126, 133; *Proctor* 9649; *Weiss* 169-72.

REFERENCES

136 *The completion* . . . *Bühler* (37) 65-7; *CP*: Essex; *Crotch* 76; *Scofield* I 397.
137 *But his* . . . *Crotch* 77-8.
138 *No doubt* . . . *Blake (World)* 92; *Kendall* 221-6, 277-8.
138 *Aesop's* . . . *Hodnett* 2; *Lenaghan* 8-9; *Wilson (Esope).*
139 *Caxton added* . . . *Blake (Prose)* 55-7.
139 *Meanwhile Richard's* . . . *Duff (Century)* xi-xii, 1, 47, 78; *Duff (West-
      minster)* 53-4, 74-7; *Kendall* 251-2, 284-5; *Kerling.*

14 ORDERS OF CHIVALRY

141 *On 1 March* . . . *Chrimes* 28-9; *Kendall* 205-7; *Myers* 340-3.
141 *In view* . . . *Chrimes* 28, 35; *Crotch* 82; *Kendall* 67, 257-8, 290; *Scofield* II
      205-6.
142 *Richard* . . . *Crotch* 83-4; *Kendall* 444; *Mitchell* 102.
142 *Caxton's censure* . . . *Crotch* 82-3.
143 *Golden* . . . *Blake (Prose)* 88-91; *Crotch* 70-3; *Hodnett* 3-4.
144 *Old William* . . . *CP*: Arundel; *Kendall* 189-90, 277, 293, 302; *Scofield* I
      140, 417, 503, 523, II 1-2, 12, 125.
145 *Golden* . . . *Blake (World)* 117-23; *Crotch* 72, 76.
145 *In Golden* . . . *Blake (Prose)* 91-6; *Crotch* 74-6.
146 *Morte* . . . *Blake (JR)* 39-40; *Blake (World)* 94-5; *Chrimes* 29, 38; *Crotch*
      92-5; *Myers* 342-3; *Kendall* 254-6; *Lander* 253-6; *Lyell-Watney* 173;
      Williams 13, 23, 26, 54.
147 *Caxton's edition* . . . *Blake (Prose)* 108-11; *Crotch* 92-3; *Vinaver* I viii.
148 *Owing* . . . *Bühler* (40); *Vinaver* I cxxx.
148 *Charles* . . . *BMC* VIII xvi, 237, 365, 368, *Herrtage* vii.
148 *Charles* . . . *Calendar (Patent)* 1476-85 223, 258, 276, 375, 387, 391-2, 433;
      Chrimes 111; *Herrtage* vii-xii; *Lander* 279; *Myers (Household)* 22,
      244-5; *Pollard, A. F.* I 100-1, 109-12, 152.
149 *Charles* . . . *BMC* IX xlii; *Duff (Westminster)* 88-91; *Leach* ix-xxxi.

15 WITH A STRANGE DEVICE

152 *Caxton's introduction* . . . *BMC* IX lvi-lvii; *HPT* I 63-7.
152 *The new* . . . *Blades* II pl. XXI.
153 *The only* . . . *Bradshaw* 84-100; *Dodgson*; *Cooke-Wordsworth* II 646-58.
154 *This Horae* . . . *Hodnett* p. 6, nos. 337-73.
154 *Speculum* . . . *Blades* II 196; *Blades (1882)* 318; *Hodnett* 5-6, nos. 309-36.
155 *The undated* . . . *Blades* II 190-1; *Blake (Prose)* 150; *Blake (World)* 87;
      Crotch xlv, cx, 99-102; *Duff (XV)* no. 248; *Gallagher*; *Plomer (Caxton)*
      149.
156 *Type 4** . . . *De Ricci* 98, 99; *Duff (XV)* nos. 408, 409; *Hellinga (Ovid)*;
      HPT I 54.
156 *Two other* . . . *Blades* II pl. XXXVII.
157 *Caxton's folio* . . . R. Proctor, letter to *The Times*, 27 Feb. 1893.
157 *It was* . . . *BMC* III 732, VIII xx, xxix; *Duff (Westminster)* 17; *Duff (XV)*
      no. 322; *Morgan-Painter.*
158 *Caxton's device* . . . *Blades* I 121; *Duff (XV)* no. 247; *Duff (Century)* 103;
      Duff (Westminster) 18; *Duff (WC)* 70; *Morgan-Painter*; *Tanner* 165;
      Walcott 140-1.
160 *It is* . . . *Blake (World)* 35, 76; *Kerling*; *Morgan-Painter* 232.
160 *Caxton must* . . . *Morgan-Painter* 234.
160 *Caxton's device* . . . *Blades* II, pl. XVIII; *Blades (1882)* 138-9; *Countryman*
      XLIV (1951) 104, XLV (1952) 396; *Girling* 22, 24, 117; *VCH Hertford-
      shire* III 365.
161 *Crotch* . . . *Cousland*; *Crotch* xlvi-xlvii; *McCusker*; *Stowe*, Charters nos. 129,
      130.
161 *What of* . . . *Crotch* xlvi-xlvii.

WILLIAM CAXTON

16 THE TWO QUEEN MOTHERS

163 *No known* . . . *Blades* II pl. LI, 221.
163 *For Reynard* . . . *Blake* (*Reynard*) 74/5, 94/22; *Chrimes* 71, 88 n. 2; *Crotch* 63; *Emden* (*Oxford*): Sant; *Lander* 275; *OED*: growl; *Pollard, A. F.*, I 43, 45, 84–7; *Rot. Parl.* VI 436–7; *Williams, N.* 63; *Williams, C. H.*; *Wedgewood* 792–3.
164 *The King* . . . *Chrimes* 54–5, 92; *Crotch* 106–7; *CP: Oxford*; *Duff* (*WC*) 84; *OED: Bayard*; *Scofield* II 190-1, 213-14.
166 *At the* . . . *Anglo*; *Crotch* 104–5.
167 *Early* . . . *Calendar* (*Venice*) I 553; *Chrimes* 62, 76; *Lander* 270; *MacGibbon* 190, 196; *Mackie* 69, 87; *Williams, N.* 37, 60.
168 *Another* . . . *Chrimes* 330; *Pollard, A. F.* I 28-9; *Weiss* (*Gigli*); *Calendar* (*Venice*) nos. 548, 550, 551, 553.
168 *The Indulgence* . . . *Barker*; *Calendar* (*Close*) Henry VII no. 203; *Clive* 241-2, 286-7; *Doyle* 300; *Kendall* 147, 324, 471; *Lyte* 72-3, 83-4; *Pollard, A. F.* I 196-7; *Scofield* II 434-8.
169 *Meanwhile* . . . *Byles* xii, xvii, 16-18; *Chrimes* 280-1; *Crotch* 103-4; *Lander* 274-5; *Mackie* 91-103.
170 *Instead* . . . *Blades* I 34, 72, II 205; *Byles* xxix-xxx.
170 *Doctrinal* . . . *Blades* II 200; *Gallagher.*
172 *A second* . . . *Beaven* I 168, II 15; *Blades* II pl. LV.

17 THE ART OF DYING

173 *Thanks to* . . . *Bean* 242-8; *Blades* II 211-12; *Chrimes* 182; *Pollard, A. F.* III 320.
173 *In 1490* . . . *Blake* (*Prose*) 157; *Crotch* 5-6, 96, 103-5, 107-10; *Stow* II 105, 375-6.
176 *Lastly* . . . *Crotch* 109; *Duff* (*WC*) 77; *Edwards* 36; *Pollard, A. F.* I 73; *Pollet* 12.
177 *Caxton's friendly* . . . *Edwards* 54; *Pollard, A. F.* I 299; *Pollet* 19, 21, 38; *Salter.*
177 *Caxton completed* . . . *Blades* II 198; *Scholderer* (*Wales*)34, no. 116.
178 *Three* . . . *Blades* II pl. XLIX; *Duff* (*Comm*); *Pfaff* 97-103.
178 *Entries* . . . *Chrimes* 280-1; *Crotch* cxxiii, clvii-clviii; *Mackie* 102-3.
179 *At the same* . . . *Crotch* cxxv; *Letts* (*M*) 17, 133; *Pollard, A. F.* III 223-9.
179 *The churchwardens'* . . . *Blades* I 70, 120; *Blades* (1882) 31, 81; *Crotch* lxxxi, cxxii; *Duff* (*WC*) 88.
180 *Caxton translated* . . . *GW* 2584, 2615, 2634.
180 *Soon after* . . . *Aurner* 137-9; *MacCracken* IXv.
181 *Caxton's new* . . . *Blades* (1882) pl. XVIII; *BMC* IX 39; *Duff* (*XV*) pp. 126-7.
182 *Book of* . . . *Blake* (*Prose*) 101-2.
183 *Ars* . . . *Aurner* 107; *Blades* (1882) 359.
183 *Caxton's last* . . . *Crotch* 111; *Hodnett* 6; *Meier-Ewart.*
184 *A copy* . . . *Bradshaw* 341-9; *Painter* (*Caxton*); *Tanner* 158-9.
186 *Now that* . . . *Croft*; *Crotch* 70; *Plomer* (*Worde*) 52.
187 *Meanwhile* . . . *BMC* VIII xlvi; *Crotch* 7; *Duff* (*XV*) no. 235.
187 *Wynkyn's* . . . *Duff* (*XV*) no. 235.
187 *Caxton last* . . . *Crotch* clii, clvii; *Pearce* 154-5, 163-4; *Scott* 1900.
188 *The biennial* . . . *Blades* I 75, 120; *Crotch* cxxiv; *Duff* (*Century*) 25; *Duff* (*WC*) 86; *Walcott* 140; *Westlake* (*St M*) 155.
188 *The eighteenth* . . . *Ames-Herbert* I 108; *Blades* I 75; *Crotch* cxxiv; *Duff* (*Century*) 25; *Duff* (*WC*) 86-7; *STC* 9995-10000.
188 *Caxton's will* . . . *Duff* (*Century*) 173; *Lauritis* 1; *Walcott* 140.
189 *Caxton evidently* . . . *BMC* IX lix, 197; *HPT* I 71-3, 92; *Duff* (*Westminster*) 24; *Plomer* (*Worde*) 45-7.
190 *One man* . . . *Crotch* cxxvii-cxxix, clx-clxiii; *Tanner* 155.
191 *So the* . . . *Lander* 283; *Pollard, A. F.* I xxii, 103, 196, 199.

CHRONOLOGICAL LIST OF CAXTON'S EDITIONS

All known Caxton editions are here listed by title, as nearly as possible in chronological order, in accordance with the evidence and arguments brought forward in this book. Undated works are supplied with inferential dates enclosed in square brackets. Each entry includes basic bibliographical particulars comprising format, number of leaves, types used, and (where available) references to E. G. Duff, *Fifteenth Century English Books* (1917), S. de Ricci, *Census of Caxtons* (1909), F. R. Goff, *Incunabula in American Libraries. A third census* (1964), and *Short-title Catalogue of Books printed in England 1475-1640* (1926), or *STC*. *Duff* provides full bibliographical descriptions of each edition, while *De Ricci*, *Goff*, and *STC* give locations of surviving copies, with other useful information. Matters of authorship, textual history, and date of printing are discussed in the main text above. Modern editions, reprints, and facsimiles of Caxton texts are listed in N. F. Blake, *Caxton and his World* (1969), pp. 224-39, *Cambridge Bibliography of English Literature*, vol. 1 (1974), cols. 667-74, and W. L. Heilbronner, *Printing and the Book in Fifteenth-Century England* (1967).

BRUGES

Recuyell of the Histories of Troy. [1475.] Folio. 352 leaves. Type 1. *Duff* 242; *De Ricci* 3; *Goff* L-117; *STC* 15375.
Game of Chess I. [After 31 March 1475.] Folio. 74 leaves. Type 1. *Duff* 81; *De Ricci* 1; *Goff* C-413; *STC* 4920.
Recueil des histoires de Troie. [1475.] Folio. 286 leaves. Type 1. *Duff* 243; *De Ricci* 3b; *Goff* L-113.
Méditations sur les sept Psaumes pénitentiaux. [1475.] Folio. 34 leaves. Type 1. *Duff* 25; *De Ricci* 3d.
Histoire de Jason. [1476.] Folio. 134 leaves. Type 1. *Duff* 244; *De Ricci* 3c.
Cordiale. [1476.] Folio. 74 leaves. Type 2. *Duff* 108; *De Ricci* 2; *Goff* C-908.

WESTMINSTER

Indulgence; commissary: John Sant. [Not after 13 December 1476.] Single half leaf. Types 2, 3. See *Pollard, A. W.* (*Ind*).
Cato I. [1476.] 4º. 34 leaves. Type 2. *Duff* 76; *De Ricci* 13; *STC* 4850.
Churl and Bird I. [1476.] 4º. 10 leaves. Type 2. *Duff* 256; *De Ricci* 66; *STC* 17008.
Horse, Sheep and Goose I. [1476.] 4º. 18 leaves. Type 2. *Duff* 261; *De Ricci* 69; *STC* 17018.
History of Jason. [1477.] Folio. 150 leaves. Type 2. *Duff* 245; *De Ricci* 64; *Goff* L-112; *STC* 15383.
Dicts of the Philosophers I. 18 November 1477. Folio. 78 leaves. Type 2. *Duff* 123; *De Ricci* 36, 37; *Goff* D-272; *STC* 6826.
Moral Proverbs. 20 February 1478. Folio. 4 leaves. Type 2. *Duff* 95; *De Ricci* 27; *Goff* C-473; *STC* 7273.
Cato II. [1477 or 1478.] 4º. 34 leaves. Type 2. *Duff* 77; *De Ricci* 14; *Goff* C-314; *STC* 4851.
Parliament of Fowls. [1477 or 1478.] 4º. 24 leaves. Type 2. *Duff* 93; *De Ricci* 25; *STC* 5091.

Anelida and Arcite. [1477 or 1478.] 4º. 10 leaves. Type 2. *Duff* 92; *De Ricci* 24; *STC* 5090.

Book of Courtesy. [1477 or 1478.] 4º. 14 leaves. Type 2. *Duff* 53; *De Ricci* 11; *STC* 3303.

Churl and Bird II. [1477 or 1478.] 4º. 10 leaves. Type 2. *Duff* 257; *De Ricci* 67; *Goff* L-406; *STC* 17009.

Horse, Sheep and Goose II. [1477 or 1478.] 4º. 18 leaves. Type 2. *Duff* 262; *De Ricci* 70; *Goff* L-407; *STC* 17019.

Stans puer ad mensam. [1477 or 1478.] 4º. 4 leaves. Type 2. *Duff* 269; *De Ricci* 74; *Goff* L-411; *STC* 17030.

Temple of Glass. [1477 or 1478.] 4º. 34 leaves. Type 2. *Duff* 270; *De Ricci* 75; *STC* 17032.

Horae ad usum Sarum I. [1477 or 1478.] 8º. (?) leaves. Type 2. *Duff* 174; *De Ricci* 50; *Goff* H-420; *STC* 15867.

Infantia Salvatoris. [1477 or 1478.] 4º. 18 leaves. Type 2. *Duff* 222; *De Ricci* 62; *Goff* I-73; *STC* 14551.

Propositio Johannis Russell. [1477 or 1478.] 4º. 4 leaves. Type 2. *Duff* 367; *De Ricci* 90; *STC* 21458.

Canterbury Tales I. [1478.] Folio. 374 leaves. Type 2. *Duff* 87; *De Ricci* 22; *Goff* C-431; *STC* 5082.

Boethius. [1478.] Folio. 94 leaves. Types 2, 3. *Duff* 47; *De Ricci* 8; *Goff* B-813; *STC* 3199.

Cordiale. 24 March 1479. Folio. 78 leaves. Types 2*, 3. *Duff* 109; *De Ricci* 33; *Goff* C-907; *STC* 5758.

Nova Rhetorica. [1479.] Folio. 124 leaves. Type 2*. *Duff* 368; *De Ricci* 91; *STC* 24189.

Ordinale ad usum Sarum. [1479.] 4º. (c. 130?) leaves. Type 3. *Duff* 336; *De Ricci* 82; *STC* 16228.

Advertisement. [1479.] Single half leaf. Type 3. *Duff* 80; *De Ricci* 17; *STC* 4890.

Horae ad usum Sarum II. [1479.] 4º. (?) leaves. Type 3. *Duff* 175; *De Ricci* 51; *STC* 15868.

Dicts of the Philosophers II. [1480.] Folio. 78 leaves. Type 2*. *Duff* 124; *De Ricci* 38; *Goff* D-273; *STC* 6828.

Epitome Margaritae eloquentiae. [1480.] Folio. 34 leaves. Type 2*. *STC* 24190.3. See *Mortimer.*

Indulgence; commissary: John Kendale. [Not after 31 March 1480.] Single half leaf. Type 2*. *Duff* 204; *De Ricci* 56; *STC* 22582.

Officium Visitationis B.V.M. [1480.] 4º. (24?) leaves. Type 4. *Duff* 148; *De Ricci* 43; *STC* 15848.

Doctrine to learn French and English. [1480.] Folio. 26 leaves. Type 4. *Duff* 405; *De Ricci* 97; *Goff* V-315; *STC* 24865.

Chronicles of England I. 10 June 1480. Folio. 182 leaves. Type 4. *Duff* 97; *De Ricci* 29; *Goff* C-477; *STC* 9991.

Description of Britain. 18 August 1480. Folio. 30 leaves. Type 4. *Duff* 113; *De Ricci* 35; *Goff* C-477(2); *STC* 13440a.

Indulgence; commissary John Kendale. [Not before 7 August 1480.] Single half leaf. Type 4. *Duff* 207; *De Ricci* 57; *STC* 22584.

Mirror of the World I. [After 8 March 1481.] Folio. 100 leaves. Type 2*. *Duff* 401; *De Ricci* 94; *Goff* M-883; *STC* 24762.

Reynard the Fox I. [After 6 June 1481.] Folio. 85 leaves. Type 2*. *Duff* 358; *De Ricci* 87; *Goff* R-137; *STC* 20919.

Of Old Age, Of Friendship; Of Nobility. 12 August 1481. Folio. 120 leaves. Types 2*, 3. *Duff* 103; *De Ricci* 31; *Goff* C-627; *STC* 5293.

Indulgence; commissary: Johannes de Gigliis. Single issue. [Not before 7 August 1481.] Single half leaf. Type 4. *Duff* 209; *De Ricci* 58; *Goff* S-565; *STC* 22586.

Indulgence; commissary: Johannes de Gigliis. Plural issue. [Not before 7 August 1481.] Single half leaf. Type 4. *Duff* 210; *De Ricci* 59; *STC* 22587.

Godfrey of Boloyne. 20 November 1481. Folio. 144 leaves. Type 4. *Duff* 164; *De Ricci* 46; *Goff* G-316; *STC* 13175.

Chronicles of England II. 8 October 1482. Folio. 182 leaves. Type 4. *Duff* 98; *De Ricci* 30; *Goff* C-478; *STC* 9992.

Polycronicon. [After 2 July 1482, before 20 November 1482.] Folio. 450 leaves. Type 4. *Duff* 172; *De Ricci* 49; *Goff* H-267; *STC* 13438.

Cato III. [1482.] Folio. 28 leaves. Types 2*, 3. *Duff* 78; *De Ricci* 15; *STC* 4852.

Game of Chess II. [1482.] Folio. 84 leaves. Type 2*. *Duff* 82; *De Ricci* 18; *Goff* C-414; *STC* 4921.

Psalterium. [1483.] 4o. 177 leaves. Type 3. *Duff* 354; *De Ricci* 84; *STC* 16253.

Curia sapientiae. [1483.] Folio. 40 leaves. Type 4. *Duff* 260; *De Ricci* 68; *STC* 17015.

Pilgrimage of the Soul. 6 June 1483. Folio. 114 leaves. Type 4. *Duff* 267; *De Ricci* 73; *Goff* G-640; *STC* 6473.

Festial. 30 June 1483. Folio. 116 leaves. Type 4*. *Duff* 298; *De Ricci* 79; *Goff* M-620; *STC* 17957 (1).

Quattuor sermones I. [July 1483.] Folio. 30 leaves. Type 4*. See *Webb.*

Confessio amantis. 2 September 1483. Folio. 222 leaves. Types 4, 4*. *Duff* 166; *De Ricci* 48; *Goff* G-329; *STC* 12142.

Curial. [1483.] Folio. 6 leaves. Type 4*. *Duff* 84; *De Ricci* 20; *Goff* C-429; *STC* 5057.

Canterbury Tales II. [1483.] Folio. 312 leaves. Types 2*, 4*. *Duff* 88; *De Ricci* 23; *Goff* C-432; *STC* 5083.

Book of Fame. [1483.] Folio. 30 leaves. Type 4*. *Duff* 86; *De Ricci* 21; *STC* 5087.

Troilus and Criseyde. [1483.] Folio. 120 leaves. Type 4*. *Duff* 94; *De Ricci* 26; *STC* 5094.

Life of Our Lady. [1483.] Folio. 96 leaves. Type 4*. *Duff* 266, 266a; *De Ricci* 71, 72; *Goff* L-409; *STC* 17023.

Sex epistolae. [1483.] 4o. 24 leaves. Types 3, 4*. *Duff* 371; *De Ricci* 92; *STC* 22588.

Knight of the Tower. 31 January 1484. Folio. 106 leaves. Types 4, 4*. *Duff* 241; *De Ricci* 63; *Goff* L-72; *STC* 15296.

Cato IV. [After 23 December 1483 (c. February 1484).] Folio. 80 leaves. Types 2*, 4*. *Duff* 79; *De Ricci* 16; *Goff* C-313; *STC* 4853.

Quattuor sermones II. [(c. February)1484.] Folio. 30 leaves. Type 4*. *Duff* 299; *De Ricci* 85; *Goff* Q-14; *STC* 17957(2).

Aesop. 26 March 1484. Folio. 144 leaves. Types 3, 4*. *Duff* 4; *De Ricci* 4; *STC* 175.

Order of Chivalry. [(c. April) 1484.] 4o. 52 leaves. Types 3, 4*. *Duff* 58; *De Ricci* 81; *Goff* O-93; *STC* 3326.

Golden Legend. [1484.] Folio. 449 leaves. Types 3, 4*. *Duff* 408; *De Ricci* 98; *Goff* J-148; *STC* 24873.

Morte d'Arthur. 31 July 1485. Folio. 432 leaves. Type 4*. *Duff* 283; *De Ricci* 76; *Goff* M-103; *STC* 801.

Charles the Great. 1 December 1485. Folio. 96 leaves. Type 4*. *Duff* 83; *De Ricci* 19; *STC* 5013.

Paris and Vienne. 19 December 1485. Folio. 36 leaves. Type 4*. *Duff* 337; *De Ricci* 83; *STC* 19206.

Directorium sacerdotum I. [1486.] Folio. 160 leaves. Type 5. *Duff* 290; *De Ricci* 77; *STC* 17720.

Image of Pity I. [1486.] Single folio leaf. Woodcut. *De Ricci* 54; *STC* 14072.

Horae ad usum Sarum III. [1486.] 8o. (?) leaves. Type 5. *Duff* 178; *De Ricci* 52; *STC* 15871.

Speculum vitae Christi I. [1486.] Folio. 148 leaves. Type 5. *Duff* 48; *De Ricci* 9; *STC* 3259.

Royal Book. [1487.] Folio. 162 leaves. Type 5. *Duff* 366; *De Ricci* 89; *Goff* L-91; *STC* 21429.

WILLIAM CAXTON

Golden Legend. [1487.] Folio. Types 4*, 5. *Duff* 409; *De Ricci* 99; *Goff* J-149; *STC* 24874. *N.B. Not entitled to the status of a second or separate edition, being a reprint of 256 of 448 leaves to make up deficiencies in the first edition.*

Deathbed Prayers. [1487? or 1485?] Single folio leaf. Types 3, 4*. *Duff* 112; *De Ricci* 34; *STC* 6442=14554.

Life of St. Winifred. [1487? or 1485?] Folio. 16 leaves. Type 4*. *Duff* 414; *De Ricci* 100; *Goff* W=62; *STC* 25853.

Donatus. [1487.] Folio. (?) leaves. Type 5. *Duff* 129; *De Ricci* 41; *STC* 7013.

Book of Good Manners. 11 May 1487. Folio. 66 leaves. Type 5. *Duff* 248; *De Ricci* 65; *STC* 15394.

Missale ad usum Sarum. 4 December 1487. Printed for Caxton by Guillaume Maynal, Paris. Folio. 266 leaves. *Duff* 322; *De Ricci* 102; *STC* 16164.

Legenda ad usum Sarum. 14 August 1488. Printed for Caxton by Guillaume Maynyal, Paris. Folio. 372 leaves. *Duff* 247; *De Ricci* 101; *STC* 16136. See *Morgan-Painter.*

Reynard the Fox II. [1488.] Folio. 72 leaves. Type 6. *Duff* 359; *De Ricci* 88; *STC* 20920.

Directorium sacerdotum II. [1488.] Folio. 194 leaves. Types 4*, 6. *Duff* 292; *De Ricci* 78; *STC* 17722.

Four Sons of Aymon. [1488.] Folio. 278 leaves. Type 6. *Duff* 152; *De Ricci* 45; *STC* 1007.

Blanchardin and Eglantine. [1488.] Folio. 98 (+) leaves. Type 6. *Duff* 45; *De Ricci* 7; *STC* 3124.

Dicts of the Philosophers III. [1489.] Folio. 70 leaves. Type 6. *Duff* 125; *De Ricci* 39; *Goff* D-274; *STC* 6829.

Indulgence; commissaries: Johannes de Gigliis and Perseus de Malviciis. [Not after 24 April 1489.] Single half leaf. Type 7. *Duff* 211; *De Ricci* 60; *STC* 14100. *N.B. Not a 'singular issue' as Duff says, as blanks are left to be filled in by hand to suit one or more purchasers.*

—, —. *Duff* 212; *De Ricci* 61; *STC* 14101. *N.B. This is a true 'singular issue'.*

Faytes of Arms. 14 July 1489. Folio. 144 leaves. Type 6. *Duff* 96; *De Ricci* 28; *Goff* C-472; *STC* 7269.

Doctrinal of Sapience. [After 7 May 1489.] Folio. 92 leaves. Type 5. *Duff* 127; *De Ricci* 40; *Goff* D-302; *STC* 21431.

Mirror of the World II. [1489.] Folio. 88 leaves. Type 6. *Duff* 402; *De Ricci* 95; *Goff* M-884; *STC* 24763.

Statutes 1, 3, 4 Henry VII. [1490.] Folio. 42 leaves. Type 6. *Duff* 380; *De Ricci* 93; *STC* 9348.

Eneydos. [After 22 June 1490.] Folio. 86 leaves. Type 6. *Duff* 404; *De Ricci* 96; *Goff* V-199; *STC* 24796.

Speculum vitae Christi II. [1490.] Folio. 148 leaves. Type 5. *Duff* 49; *De Ricci* 10; *Goff* B-903; *STC* 3260.

Horae ad usum Sarum IV. [1490.] 8o. (?) leaves. Type 5. *Duff* 179; *De Ricci* 53; *STC* 15872.

Officium Transfigurationis Jesu Christi. [1490.] 4o. 10 leaves. Type 5. *Duff* 146; *De Ricci* 42; *STC* 15854.

Commemoratio Lamentationis B.V.M. [1490.] 4o. 32 leaves. Type 5. *Duff* 105; *De Ricci* 32; *STC* 17534.

Image of Pity II. [1490?] Single leaf quarto. Woodcut. *De Ricci* 55; *STC* 14072a

Art to know well to die. [After 15 June 1490.] Folio. 14 leaves. Type 6. *Duff* 35; *De Ricci* 6; *STC* 789.

Governal of Health. [1491.] 4o. 18 leaves. Type 6. *Duff* 165; *De Ricci* 47; *Goff* G-328; *STC* 12138.

Book of Diverse Ghostly Matters. [1491.] 4o. 148 leaves. Type 6. *Duff* 55; *De Ricci* 12; *Goff* G-301; *STC* 3305.

Ars moriendi. [1491.] 4o. 8 leaves. Types 6, 8. *Duff* 33; *De Ricci* 5; *STC* 786.

Festial II. [1491.] Folio. 136 leaves. Types 6, 8. *Duff* 301; *De Ricci* 80; *Goff* M-621; *STC* 17959 (1).

CAXTON'S DEVICE FOR DATING

Quattuor sermones III. [1491.] Folio. 34 leaves. Type 6. *Duff* 302; *De Ricci* 86; *Goff* Q-15; *STC* 17959(2).
Fifteen Oes. [1491.] 4º. 22 leaves. Type 6. *Duff* 150; *De Ricci* 44; *STC* 20195.
Horae ad usum Sarum V. [1491.] 8º. (?) leaves. Type 8.
Horae ad usum Sarum VI. [1491.] 4º. (?) leaves. Type 8.

Caxton also perhaps printed editions of his own translations of Ovid, *Metamorphoses moralised*, completed 22 April 1480, and of *Life of Robert Earl of Oxford*, written towards the beginning of 1488. No copy of either is known to survive, but their existence seems more likely than not.

CAXTON'S DEVICE AS EVIDENCE FOR DATING

Progressive damage to the bottom frameline of Caxton's device occurs in six stages, here classified as States A-F, and so supplies evidence for inferential year dates and chronological order in respect of the ten undated works in which the device is used.

STATE A (1487)

With a 10 mm. break in bottom frameline at extreme right:
Missale ad usum Sarum, 4 December 1487

STATE B

With a second 10 mm. break 6 mm. to left of first:
Legenda ad usum Sarum, 14 August 1488
Reynard the Fox II
Directorium sacerdotum II

STATE C (early 1489)

With a third break of 6 mm., 10 mm. to left of second break:
Dicts of the Philosophers III

STATE D (later 1489)

With the previous breaks joined and continuous, slightly lengthened, totalling 44 mm.:
Doctrinal of Sapience after 7 May 1489
Mirror of the World II

STATE E (1490)

With the break enlarged to 70 mm., but with traces of the broken portion still visible:
Speculum vitae Christi II
Eneydos after 22 June 1490

STATE F (1491)

As State E, but with a new break in left frameline near foot:
Book of Diverse Ghostly Matters
Festial II
Quattuor sermones III

INDEX

Abingdon, Benedictine abbey 84, 157, 164, 183
Abusé en cour 80 *n*3, 137 *n*2
Actors, Peter 116 *n*3, 140 *n*3
Ailly, Pierre d' 78; *Jardin de dévotion* 72-3, 81; *Méditations sur les sept Psaumes pénitentiaux: see* Caxton
Aleyn, Richard 98
Alfonso, King of Naples 115 *n*1
Alost 61*n*
Ames, Joseph 94, 188
Anne, Duchess of Brittany 179; of York, d. of Edward IV 119, 128
Antwerp 21, 24, 26, 28, 35 *n*1, 36-8, 74 *n*2, 145-6, 150-4, 189
Anwykyll, John 136 *n*1, 140 *n*2
Apollonius of Tyre 84
Armagnac, Count of 51 *n*2
Arques 51 *n*3
Ars moriendi 80 *n*3, 180, 183; *see also* Caxton: *Ars moriendi*; *Art to know well to die*
Arthur, King 37, 41, 115, 147-8, 166-167; *see also* Caxton: *Morte d'Arthur*; Prince, s. of Henry VII 135, 147, 166-7, 173, 177
Arundel, Thomas, Archbishop 154; Thomas, Prior 100; William Fitz-Alan, Earl of 116, 144-5, 152, 165, 187
Aubriet, Guillaume 21 *n*3
Aucassin et Nicolette 150
Aureus de Universo 117
Aurner, Nellie S. 125 *n*1
Austrey, Warw. 13 *n*2

Bagnyon, Jean 148
Ballard, Mr, of Chipping Campden 188
Bamberg 54*n*
Bamborough Castle 30 *n*3
Banste, Sir John de 110
Barker, Nicolas 126 *n*2
Barking, Essex 37 *n*2
Barnet, battle of 46-7, 65, 86, 165
Bartholomaeus Anglicus, *De proprietatibus rerum* 54-9, 74-5, 118
Basel 53, 56-7, 158 *n*1 & 2; University Library 56
Bayard, the magic horse 165
Beauchamp, Richard, Bishop of Salisbury 35
Bechtermünzer, Heinrich 54*n*

Bergen, Henry of 191
Bergen-op-zoom 21, 27*n*
Berkeley, Thomas, Earl of 117
Bermondsey Abbey 167, 186, 191
Berry, Jean, duc de 155
Bersuire, Pierre 101*n*
Biblioteca Colombina, Seville 78 *n*3
Bibliothèque Nationale, Paris 80
Bircholt, Kent 4, 5
Black Book of the King's Household 149 *n*1
Blades, William 5 *n*1, 8, 9 *n*2, 10, 11 *n*2, 12 *n*1, 13 *n*2, 14 *n*1, 17 *n*2, 18, 19, 22 *n*1, 28 *n*1 & 2, 29, 34 *n*1, 37 *n*2, 52, 65 *n*1 & 4, 73, 76 *n*3, 6 & 7, 87 *n*2, 93-5, 99, 103 *n*2, 105, 108, 119 *n*1, 120, 123, 124 *n*3, 125 *n*1, 131, 134 *n*1, 136, 154 *n*3 & 4, 155 *n*1, 156, 160 *n*3, 163*n*, 169 *n*2, 170 *n*1 & 2, 177*n*, 178 *n*2, 180, 182 *n*2, 183, 184 *n*2, 186, 188
Blake, Norman F. 8, 9, 15 *n*1, 19, 20, 21 *n*2, 26 *n*1, 45, 72, 91, 103 *n*1, 117 *n*3, 126, 128, 131, 138, 147, 161 *n*2, 176 *n*1
Blount, Sir James 166
Blavis, Bartholomaeus de 103*n*
Boccaccio, *De casibus virorum illustrium* 64, 72, 76 *n*2, 81, 174; *Genealogia deorum* 60, 76 *n*4, 86
Boethius, *De consolatione philosophiae* 74, 76 *n*2, 80 *n*3, 81*n*; *see also* Caxton: *Boethius*
Bomsted, Henry 6*n*, 27, 34
Bona of Savoy 32, 33
Bonaventure, Saint 154
Bonet, Honoré 169
Bonhomme, Pasquier 62 *n*2, 78 *n*2
Bonifaunt, Richard 10, 11, 14, 18
Book of Hawking, Hunting, and Blasing of Arms 134, 157 *n*4, 182 *n*2
Bookbinder, James 82, 184 *n*3
Bost, Henry 168
Boston, Mass., Public Library 161
Bosworth, battle of 65, 151, 157, 165
Bourchier, Sir Thomas 47 *n*1; Thomas, Cardinal 103; Sir William 93 *n*3, 137
Brackenbury, Sir Robert 135, 149
Bradley, Henry 103 *n*2
Bradshaw, Henry 186

217

Braziller, George 102 *n*2
Bremen 101
Bretailles, Louis de 87, 89
Bridget of York, d. of Edward IV
 119 *n*3, 128; Saint, of Sweden 184
Brie, F. W. D. 106 *n*1
Bristol 38 *n*1
British Library 80, 92 *n*4, 99, 104,
 106 *n*1, 108-9, 112-13, 117 *n*2,
 124 *n*1, 126 *n*1, 132, 134 *n*2,
 136 *n*2, 143 *n*3, 145 *n*1, 169 *n*2,
 178 *n*2, 179, 181, 186
British Museum *see* British Library
Brito, Jean 80
Brothers of the Common Life,
 Brussels 61
Brotherton Library, Leeds University
 104 *n*4
Bruges 8, 14-81, 83, 85-7, 94, 102,
 103, 108-10, 119, 161, 163,
 168 *n*1; *see also* Mansion, Colard
Brussels 48, 51 *n*3, 61, 81, 146
Brutus the Trojan 52, 105, 147
Bryce, Hugh 37-8, 108, 110, 172,
 187; Thomas 37 *n*2
Buckingham, Henry Stafford, Duke of
 116, 122 *n*2, 123*n*, 129, 138, 141
Bühler, Curt F. 112, 137 *n*1
Bulle, Johannes 101
Bumpstead, Richard 6*n*
Burdett, Thomas 65 *n*3, 90
Burgh, Benedict 6*n*, 93-4, 137; Richard
 6*n*, 18, 19; Robert 6*n*
Burgundy, Anthony, Bastard of 114*n*
Butler, Lady Eleanor 122

'Cacston', Thomas *see* Sawston
Cade, Jack 16, 47, 65
Cadwallader 147
Caelwaert, Joris 72 *n*2
Caerleon-on-Usk 147*n*
Caillaut, Antoine 181
Calais 5, 10, 11, 16, 20, 24-7, 30 *n*3,
 32*n*, 33, 34, 37, 40, 42, 69, 83, 86,
 108, 112 *n*3, 160 *n*3, 161-2, 179
Calixtus III, Pope 178
Cambrai 78
Cambridge 93, 96; University Library
 35, 154 *n*1 & 5, 158, 186
Cantelow, William 41
Canterbury 4, 47, 49, 116
Capons, Sir John 146
Carmelianus, Petrus 135-6, 140 *n*2,
 166-7
Caston, William, of Calais 20, 161-2
Catesby, William 129
Catherine of York, d. of Edward IV
 119 *n*3, 128

Cato, *Disticha moralia, see* Caxton:
 Cato
Cattlyn, John 105 *n*; Richard 105*n*
Caulibus, Johannes de 154
Causton, Alexander de 3; Henry de 3;
 Hugh de 4; John de 3; Michael
 de 3; Nicholas de 3; Oliver 3;
 Richard de 3; Roger de 3, 4;
 Stephen 4; Theobald de 3;
 Thomas *see* Sawston; Walter 4;
 William (1401) 3; William
 (1406) 3, 5; William, tailor 3;
 William de (1297) 3; William
 de (1354) 3, 5
Causton Wood, Kent 4 *n*1, 7
Caustons, manor, Kent 4, 6
Cawston, John, of Norwich 6; Roger
 de, of Norwich 6
Cawston, Norf. 2, 3, 7; Shrops. 2 *n*1;
 Warw. 2 *n*1
Caxton, Denise, of Little Wratting 6;
 Elizabeth, the Printer's d. 100,
 190-1; Hugh, of Sandwich 4,
 5 *n*2, 47; John, of Canterbury 4;
 John, of Hadlow Hall, Essex 6;
 John, of Westminster 6, 83;
 Matilda de 3; Maud, the Printer's
 wife (?) 88 *n*2, 100, 179-80, 188;
 Oliver, skinner 2 *n*2, 5, 180;
 Philip, attorney 6*n*; Philip, of
 Little Wratting, father 6; Philip,
 of Little Wratting, son 6; Richard,
 mercer 3, 5 *n*2, 9 *n*2, 21 *n*1;
 Richard, monk 2 *n*2, 5, 83, 100;
 Robert, of Canterbury 2 *n*2, 4;
 Thomas I, of Tenterden 2 *n*2,
 4-5, 7, 83; Thomas II, of Tenter-
 den 5, 7, 47; William, of Canter-
 bury 4; William, of Little Wrat-
 ting 2 *n*2, 6; William, of Sham-
 well Hundred 4; William, of
 Westerham 4; William, of West-
 minster 2 *n*2, 6, 83, 180; William
 de (1311) 3; William de
 (1444) 3
CAXTON, William, the Printer:
 Advertisement 98-101, 120
 Aesop 105, 130, 137-9, 143, 144,
 180 *n*3
 Anelida and Arcite 93, 95
 Ars moriendi 182-3
 Art to know well to die 80 *n*3, 151 *n*2,
 180, 183
 Blanchardin and Eglantine 166-7,
 174
 Boethius 80 *n*3, 92-3, 95 *n*2, 98, 99,
 120, 155 *n*4
 Book of Courtesy 93, 95, 189

CAXTON—continued

Book of Diverse Ghostly Matters
182-3
Book of Fame 131-2, 135
Book of Good Manners 92, 151,
155, 170, 172
Canterbury Tales I 74, 85, 91, 107,
132, 135; II 70 n2, 91, 131-5, 139
Cato I 6n, 80 n3, 93, 119; II 93,
119; III 93, 118-19; IV 136-9, 142
Charles the Great 1, 115, 148-9, 152,
174
Chronicles of England I 12, 101-2,
105-6, 117, 118 n1, 150, 179, 188,
189; II 106 n1, 107, 116-18
Churl and the Bird I 85; II 93
Commemoratio Lamentationis BVM
136 n2, 178
Confessio amantis 118 n2, 126, 130,
135, 136
Cordiale (French) 75 n2, 79, 80, 94
Cordiale (English) 77, 84, 95-6,
98, 113
Court of Sapience, see Curia
sapientiae
Curia sapientiae 118-20, 124
Curial 131
Deathbed Prayers 156-7
Description of Britain 106-7, 147 n
Dicts of the Philosophers I 74, 80 n3,
85-90, 95, 96; II 104, 120; III
167-8
Directorium sacerdotum I 99, 153,
163; II 99, 163, 168, 178 n1
Doctrinal of Sapience 151 n2, 168,
170-2, 180 n2
Doctrine to learn French and
English 103-4, 127
Donatus 157
Eneydos 151 n2, 173-7, 180, 187
Epitome Margaritae eloquentiae 104
Faytes of Arms 151 n1, 163, 169-
170, 173, 174, 180 n2
Festial I 124-5, 129-30, 146; II
182-4
Fifteen Oes 182-6
Four Sons of Aymon 102, 164-6
Game of Chess I 59, 64-72, 75 n2,
77, 78, 111, 112, 137 n4, 165; II
70, 118-19, 132, 138, 143, 160
Godfrey of Boloyne 102, 110, 114-15,
118, 122, 143, 146, 147, 149,
180 n2
Golden Legend 4, 26 n1, 74, 102,
114, 118, 124, 127, 136, 139,
143-6, 148, 150, 156, 157 n1,
187, 189; (partial reprint) 156,
167

CAXTON—continued

Governal of Health 180-1
Histoire de Jason 52, 75 n2, 78, 80,
85 n1
History of Jason 52, 74, 78, 85-7, 95,
98, 116, 122, 150, 173, 189
Horae ad usum Sarum I 120; II 120;
III 53-5, 178; IV 178, 186; V
184-6; VI 184-6
Horse, Sheep, and Goose I 85; II 93
Image of Pity I 153-4; II 154n
Indulgence (1476: Sant) 80, 83-4,
92, 104
(1480: Kendale I), 105, 115
(1480: Kendale II) 105, 115
(1481: Gigliis, single issue)
105
(1481: Gigliis, plural issue) 105
(1489: Gigliis-Malviciis, general
issue) 153, 168-9
(1489: Gigliis-Malviciis,
singular issue) 153, 168
Infantia Salvatoris 94, 136 n2
Knight of the Tower 126-30, 136,
139, 141
Legenda ad usum Sarum 158-60,
163, 181, 189-90
Life of Our Lady 123, 131-2, 189
Life of Robert Earl of Oxford 165
Life of St. Winifred 157
Méditations sur les sept Psaumes
pénitentiaux 75 n2, 78, 80
Mirror of the World I 26 n1, 102,
108-10, 114, 116, 118 n1, 119,
144; II 172, 184
Missale ad usum Sarum 158-60, 163,
181
Moral Proverbs 90-1, 95 n2, 96, 169
Morte d'Arthur 70, 91, 102, 115,
118, 124, 132, 143, 146-9, 152,
156, 157 n2, 167 n1
Nova Rhetorica 96-8, 104, 136 n2
Of Old Age, etc. 111-13, 116,
118 n1
Officium Transfigurationis Jesu
Christi 178
Officium Visitationis BVM 102-4
Order of Chivalry 141-3, 145, 146
Ordinale ad usum Sarum 98-100,
120, 153
Ovid, Metamorphoses moralised 80
n3, 98, 101-2, 106-7, 114, 126
Paris and Vienne 148-50, 152, 189
Parliament of Fowls 93, 95
Pilgrimage of the Soul 118, 123-4,
128, 129, 154 n4
Polycronicon 12, 74, 106, 111, 116-
118, 143, 147n

CAXTON—*continued*
Propositio Johannis Russell 94-5, 103 *n*1
Psalterium 120
Quattuor sermones I 125-6, 183; *II* 125-6, 137, 183; *III* 182-3, 184
Recueil des histoires de Troie 52, 75 *n*2, 78, 80, 85 *n*1, 184
Recuyell of the Histories of Troy 1, 15, 39, 45-8, 51-4, 59-64, 66, 70, 75 *n*2, 76 *n*5, 77, 78, 86, 97, 161 *n*1, 165, 170, 174, 176*n*, 187
Reynard the Fox I 110-11, 118 *n*1, 120, 150, 180 *n*2; *II* 163-4, 168
Royal Book 148, 152, 154-6, 170, 172
Sex epistolae 131, 135-6
Speculum vitae Christi I 153-5, 172, 177-8; *II* 153*n*, 177-8
Stans puer ad mensam 93
Statutes 1, 3, 4 *Henry VII* 173
Temple of Glass 93, 95
Troilus and Criseyde 131-2, 135
Tully see *Of Old Age*
Caxton, Cambs. 2
Caxton Binder 80 *n*2, 184
Cecily, Duchess of York 47 *n*1, 122; of York, d. of Edward IV 119, 128, 129
Cessolis, Jacobus de 64; *see also* Caxton: *Game of Chess*
Chablis 155 *n*3
Charety, Richard 26 *n*2
Charlemagne 115, 149, 165; *see also* Caxton: *Charles the Great*
Charles the Bold, Duke of Burgundy 30-48, 67-71, 81, 84 *n*2, 89, 93-5, 98, 122 *n*1, 146
Charles VIII, Dauphin, later King of France 116, 119 *n*3
Charles, Duke of Guienne 51 *n*2
Charolais, Count of, *see* Charles the Bold
Chartier, Alain 131; *Curial, see* Caxton; *Quadrilogue* 74, 80
Chastising of God's Children 189
Chaucer, Geoffrey 92, 131-5, 165 *n*4; *see also* Caxton: *Anelida and Arcite; Boethius; Book of Fame; Canterbury Tales; Parliament of Fowls; Troilus and Criseyde*
Chertsey 47
Childe, Alexander 161
Childe of Bristowe 14
Christine de Pisan 90-1, 169; *see also* Caxton: *Faytes of Arms; Moral Proverbs*
Church, Daniel 94

Cicero, Marcus Tullius 35, 94, 176; *see also* Caxton: *Of Old Age, etc.*
Clarence, George, Duke of 29-30, 36, 39-42, 46, 65-9, 71, 74, 81, 82 *n*2, 89, 90, 119, 122, 146 *n*3, 165 *n*5, 187, 191
Claudian 176 *n*2
Cologne 33, 35 *n*2, 38-9, 43, 46, 48-61, 67-9, 74, 76, 77, 92, 101, 118, 146
Columbus, Christopher 78, 191
Compostella 87, 89, 96
Constantinople, 52
Cook, Sir Thomas 44 *n*3
Copland, Robert 84-5, 95; William 166 *n*1
Corvo, Baron 157
Cosyn, Robert 18, 20-2
Craes, William 17
Crantz, Martin 158 *n*2
Cresse, William 162
Croft, P. J. 187
Crop, Gerard 190-1
Crotch, W. J. B. 2 *n*2, 5, 8, 9 *n*2, 10, 12 *n*1, 17 *n*2, 18, 19, 22 *n*1, 28 *n*3, 29, 37 *n*2, 51 *n*3, 53, 72 *n*2, 76 *n*7, 82 *n*3, 89 *n*3, 93 *n*2, 100 *n*2, 143 *n*3, 155 *n*4, 161-2, 178 *n*3, 190 *n*2
Cruse, Louis 148*n*
Culemborg 124 *n*4
Culham Priory 164

Daubeney, Giles, Baron 149 *n*1, 172; William 105*n*, 149, 152, 161 *n*3, 187; Sir William 149 *n*1
David, Bishop of Utrecht 33, 124 *n*4
Dedes, Robert 8, 10, 11, 14
Deguilleville, Guillaume de 123-4; *see also* Caxton: *Pilgrimage of the Soul*
Delft 68, 152
Della Porta, Roberto 72 *n*1
Della Torre, Antonio 27
De Machlinia, William *see* Machlinia
Desmond, Earl of 114*n*
De Worde, Wynkyn *see* Worde
Dialogus creaturarum 74 *n*1 & 3, 76
Dibdin, Thomas F. 94
Diodorus Siculus 176-7
Dixmude, battle of 170
Dommeloe, Hugh 35 *n*1
Donatian, St. 22
Donatus, Aelius 157; *see also* Caxton: *Donatus*
Dordrecht 68
Doria, Jacques 38
Dorset, Thomas Grey, Marquess of 121, 123, 129, 132*n*, 146, 151, 167, 168 *n*3

Dover 4, 147
Doyle, A. I. 103 *n*1
Dritzehen, Andreas 54*n*
Dudley, Jane *see* Grey
Duff, Edward G. 8, 17 *n*2, 35 *n*2, 51 *n*3, 61*n*, 84 *n*3, 94, 101, 103 *n*2, 119 *n*2, 125 *n*1, 130, 136, 154 *n*2, 155 *n*3, 156, 157 *n*4, 158, 177 *n*, 182, 188, 189
Dujardin, Simon 148 *n*
Dukmanton, Henry 11, 14
Dupré, Jean 187 *n*

Eastney, John, Abbot 82-4, 86 *n*1, 98, 123, 167, 169 *n*1, 174, 187-8
Ebesham, William 80 *n*2, 95, 103 *n*1, 112
Edgar, King 174 *n*2
Edgecot, battle of 39
Edward the Confessor, King of England 174 *n*2
Edward I 11
Edward II 3
Edward III 4, 24
Edward IV 5 *n*1, 25-51, 63-71, 81, 84 *n*1, 86, 87, 90, 98, 104-8, 114-123, 137, 141*n*, 142, 145, 149, 166, 168 *n*2, 169, 187, 191
Edward, the Black Prince 65 *n*4, 169 *n*2
Edward, Prince of Wales, later Edward V 41, 47, 65, 86-90, 96, 114-15, 119, 121-4, 129*n*, 135, 139 *n*2, 142, 149 *n*3, 167, 173
Edward, Prince of Wales, son of Henry VI 26 *n*1, 30 *n*2, 41, 46
Edward, Prince of Wales, son of Richard III 142
Edward, Earl of Warwick, son of Clarence 90*n*, 164, 166, 191
Ekwall, H. 2 *n*1
Elizabeth of York, d. of Edward IV, later Queen 116, 119, 128, 129, 141, 142*n*, 146, 149, 151, 157 *n*4, 166-8, 184, 186-7, 191
Elizabeth, Queen, wife of Edward IV, née Woodville 32, 41, 47, 63, 72, 86, 90, 114*n*, 121-4, 128-32, 137, 140-2, 146-7, 151-2, 166-8, 186, 191
Eltvil 54*n*
Empingham, battle of 41*n*
Erfurt 57
Essalen (?), Johannes 100 *n*2
Essex, Henry Bourchier, Earl of 6*n*, 93 *n*3, 137, 144 *n*1
Essex, Robert, Prior 99-100, 123, 188*n*

Eton College 168
Evilmerodach 64
Exeter, Henry Holland, Duke of 26 *n*3, 30, 36, 42, 47; Thomas, Duke of 6*n*
Eyre, Thomas 14

Falstaff, Sir John 112
Faques, William 116 *n*3
Fascet, George, Prior 188*n*
Fasciculus temporum see Rolewinck
Fastolf, Sir John 87 *n*2, 112, 114*n*
Fauconberg, Thomas, Bastard of 47, 67 *n*1, 86
Faunt, Nicholas 47
Felde, John 160 *n*3
Felding, Geoffrey 18-20, 37 *n*2
Feldman, L. D. 102 *n*2
Ferrara, war of 135-6
Ferron, Jean 64
Fierabras 148
Fillastre, Guillaume 85 *n*2
Fitzherbert, John 128 *n*1
FitzJames, Richard 117 *n*2, 190
FitzLewis, Mary 116
Flete, John, Abbot 80 *n*2, 184
Flushing 46, 48, 68
Foix, Gaston de 51 *n*2
Fotheringay 84 *n*2
Francis II, Duke of Brittany 51 *n*2
Frankenburg, Henry 140 *n*3
Frederick II, Emperor 87
Frederick III, Emperor 68, 89, 136
Friburger, Michael 158 *n*2
Froissart 143
Fructus temporum 188
Fry, Christopher 13
Fuller, Otuel 93; Thomas 4
Fust, Johann 35, 54*n*, 63

Gallet, Louis 31 *n*3
Gallopes, Jean de 124 *n*3, 154 *n*4
Gascoigne, Thomas 99
Gaunt, John of 68, 129
Gedeney, Thomas, 3, 14 *n*3
Gedney, John 14
Geneva 148
Gering, Ulrich 158 *n*2, 181
Germain, Jean 85 *n*2
Gerson, Jean 74
Ghent 21, 22, 46, 48, 51 *n*3, 53, 54, 67 *n*2, 77, 89, 94, 146
Gigliis, Johannes de 84, 104-5, 115, 166-8; Silvester de 168 *n*1
Giraldus Cambrensis 147 *n*
Glastonbury 147

Gloucester, Eleanor Cobham, Duchess of 13, 141*n*; Humphrey, Duke of 13, 16, 25, 151; Richard, Duke of *see* Richard III

Goes, Mathias van der 154 *n*1

Gops, Goiswin 57, 59 *n*3

Gossouin of Metz 108

Gouda 68, 69, 75 *n*3, 111, 160, 182

Gower, John 126, 130 *n*1, 165 *n*4; *see also* Caxton: *Confessio amantis*

Grandson, battle of 81

Granton, John 17

Grey, Lady Jane 132 *n*; Sir Richard 121 *n*3

Grierson, P. 103

Grocin, William 144 *n*3

Grosseteste, Robert 93

Gruthuyse, Louis de 22, 41, 42, 48, 64, 74, 76 *n*7

Guines 16, 30 *n*3, 33

Gutenberg, Johann 53, 54, 63, 75, 76 *n*1, 84, 97*n*

Hadlow, Kent 4, 6

Hadlow Hall, Essex 6

Hakluyt, Richard 28 *n*1

Hall, Thomas 105

Halle, Robert 12; William 12 *n*1

Hallom, Robert 176 *n*1

Hammes 90, 166

Han, Ulrich 63

Hannekin 76 *n*7

Hanseatic League 10, 17, 24, 33 *n*2, 38-42, 48-50, 54, 67-8, 95

Harleian Library 123, 136 *n*2

Harowe, John 5 *n*2, 9, 12 *n*1, 21, 22, 25

Hastings, William, Lord 41, 49, 67, 83, 87, 108, 110, 112 *n*3, 116-17, 121-2, 131 *n*1, 144, 168 *n*3, 187

Hatcliff, William 35, 37, 45, 49, 67, 68, 81

Hatfield Broadoak 165 *n*2

Hatton, Richard 190

Hawes, Stephen 119 *n*2

Haywarde, scribe 89-90

Helion Bumpstead, Essex 6*n*

Hellinga, W. and L. 70, 73, 75 *n*1

Henry II, King of England 94

Henry III 3*n*, 12 *n*2

Henry IV 24, 27, 65, 130 *n*1

Henry V 27

Henry VI 16, 25, 26 *n*1, 27, 30 *n*2, 36, 40, 41, 43, 47, 119

Henry VII 116 *n*3, 123*n*, 129, 135, 136 *n*2, 141, 142*n*, 145-52, 157, 164-73, 187, 191

Henry VIII 153, 164 *n*2, 177

Herbert, William, antiquarian 94, 105*n*; William, Earl of Pembroke 146 *n*3

Herrtage, S. J. H. 148*n*

Hesdin 85

Heton, Christopher 9-11, 14; James 12

Hexham, battle of 30 *n*3

Higden, Ranulph 106, 117; *see also* Caxton: *Description of Britain*; *Polycronicon*

Higman, Johannes 181

Hoccleve, Thomas 124

Hodnett, Edward 154 *n*2 & 5

Hoernen, Arnold ther *see* Ther Hoernen

Holinshed, Raphael 16, 105

Holywell, Flints. 157

Hopyl, Wolfgang 181

Horham manor, Essex 14 *n*1

Hornes, Philippe de 76 *n*7

Huchons, Isabella 3; Thomas 3

Hull 24, 44 *n*1

Humphrey, Good Duke *see* Gloucester

Hunte, Thomas 136 *n*1, 140 *n*3

Huntington Library, San Marino, Cal. 63, 126 *n*1

Hunyadi Janos 178

Huss, John 78

Hylton, Walter 186

Hypnerotomachia 64

Iceland 38 *n*1

Innocent VIII, Pope 136 *n*2, 168, 179

Isabel de Bourbon, 2nd wife of Charles the Bold 35, 89

Isabel of York, d. of Richard of York 137

Isabella of Portugal, wife of Philip the Good 68, 85

James III, King of Scotland 89, 116

Jenkins, manor, Barking 37 *n*2

Jermyn, Eleanor, prioress 179

John the Wise, jester 37

John Rylands Library, Manchester 89, 95, 103 *n*1, 126 *n*1, 166 *n*1

Jourdemayne, Margery 13

Kefer, Heinrich 54*n*

Kemsley, Lord 6

Kendale, John 105, 115, 136 *n*2

Kent, William Neville, Lord Fauconberg, Earl of 47 *n*1

Kerling, N. J. 160

Kingsford, C. L. 106 *n*1

Koelhoff, Johann, the Elder 57

La Cypède, Pierre de 150
Lambeth Palace Library 35, 89, 126 n1
Langley, Henry 83; Katherine 83
Langton, Thomas 136 n2
Large, Alice 11, 14; Elizabeth, née Staunton 13; Elizabeth, d. of preceding 11, 14; Johanna 11, 13, 14; John 11, 12; Richard 11, 14, 18 n1; Robert 6n, 8-15, 18n, 27n, 188; Robert, his s. 11, 14 Thomas 11, 14
La Tour Landry, Geoffroy de 126-7; *see also* Caxton: *Knight of the Tower*
Laurent, Frère 155; *see also* Caxton: *Royal Book*
Le Clerc, Jean 108
Leeu, Gerard 75 n3, 111, 150, 153, 160, 189
Lefèvre, Jean 52; Raoul 51-2, 85; *see also* Caxton: *Histoire de Jason; History of Jason; Recueil des histoires de Troie; Recuyell of the Histories of Troy.*
Legrand, Jacques 155; *see also* Caxton: *Book of Good Manners*
Le Rouge, Pierre 155 n3
Le Roy, Guillaume 78 n2, 148n, 174 n1
Lettersnider, Henrick 189 n4
Lettou, John 101, 105, 140 n1
Letts, Malcolm 23 n1
Levet, Pierre 181
Lewis, physician 129, 166 n2
Liège 179
Life of St. Katherine of Siena 189
Lille 33, 46
Lilleshall 125
Lincoln, John de la Pole, Earl of 164
Lis Belial 56, 57, 60
Litelton, Henry 18
Little Wratting, Suff. 6, 93, 137 n3
Llanegryn, Carn. 160 n3
Llanengan, Mer. 160 n3
Lokyngton, Walter 100
Losecoat Field, battle of 41n
Louis XI 32-7, 41, 44, 47 n3, 51 n2, 65, 68, 69, 83 n2, 89, 90, 98, 108
Louvain 56-61, 76, 77, 81, 117 n2, 124 n4, 152
Love, Nicholas 154
Lovel, Francis, Lord 129, 164
Ludlow 27, 96 n4, 121
Lull, Raimon 141; *see also* Caxton: *Order of Chivalry*
Lydd, Kent 5

Lydgate, John 51, 93, 119, 123-4, 165 n4, 181; *see also* Caxton: *Churl and the Bird; Governal of Health; Horse, Sheep, and Goose; Stans puer ad mensam; Temple of Glass*
Lynam, Thomas 168 n3
Lyneham, Lodovic 161
Lyons 78 n2, 137 n2, 138, 148n, 158 n2, 187n
Machlinia, William de 80, 103, 139-140, 157, 173, 180
Macho, Julien 139
Magdalene College Cambridge 102 n2; *see also* Pepysian Library
Maidstone 47
Maillet, Jacques 148n
Mainz 35, 53, 54, 63, 106 n1
Mallarmé, Stéphane 53
Malory, Sir Thomas *see* Caxton: *Morte d'Arthur*
Malviciis, Perseus de 167-8
Mancinellus, Antonius 157
Mancini, Dominico 122
Mandeville, Sir John 179
Mansion, Colard 61-4, 72-81, 85, 87, 92, 101, 102, 111n, 137 n2, 150, 156n, 160 n2, 180
Margaret Beaufort, m. of Henry VII 116 n3, 129, 141n, 166-9, 174, 184, 186-7
Margaret, Duchess of Burgundy 35-7, 42-8, 51-4, 63, 66-9, 74, 86, 89, 90, 94 n1, 98, 106, 112 n3, 116, 146 n1, 164, 165, 191
Margaret, Queen, of Anjou, wife of Henry VI 16, 25 n1, 30, 31 n3, 36, 41, 44 n3, 47, 89 n1, 105, 141n
Margaret of Scotland 89, 95, 96, 114, 116
Margate 176 n1
Mary, Duchess of Burgundy 89, 116, 146 n3; of York, d. of Edward IV 119
Master of Mary of Burgundy 72
Maximilian of Austria 89, 94 n1, 98, 116, 170
Maydestone, Clement 99, 153; *see also* Caxton: *Directorium sacerdotum*
Maynyal, Guillaume 158, 160, 163, 173, 181
Mentelin, Johann 92
Middelburg 21, 25 n2, 29, 38, 43, 68, 176 n1
Middleham 39
Miélot, Jean 72 n1, 80, 111n, 113

Milling, Thomas 82 *n*1, 86 *n*1
Milreth, William 27*n*
Mirk, John 124-5, 157; *see also* Caxton: *Festial*
Mocenigo, Giovanni, Doge of Venice 135
Montemagno, Bonaccursius de: *Controversie de noblesse* 80 *n*3, 81*n*, 111*n*; *see also* Caxton: *Of Old Age, etc.*
Montgomery, Sir Thomas 68
Monte Rocherii, Guido de 170 *n*2
Morat, battle of 81, 84 *n*2
More, Sir Thomas, Saint 168
Morgan, Paul 158
Moris, V. C. 161
Mortimer, Mrs Jean E. 104 *n*4
Morton, John 68, 179, 190-1
Mote, The, nr. Maidstone 47 *n*1, 106
Mountfort, Emma 105; Simon 105, 149
Mowbray, Anne 90
Mubashshir ibn Fatik 87

Nancy, Lorraine 89
National Library of Wales, Aberystwyth 178 *n*2
Nesfeld, John 129, 141
Neuss 69
Neve, John 21, 22, 71
Neville, Anne, wife of Richard III 41, 65, 142*n*, 146; George, Archbishop of York 31; Isabel, wife of Clarence 36, 40, 90, 146 *n*3
Newcastle-on-Tyne 24
Nieuport 19
Nixon, Howard M. 82 *n*3, 100 *n*2
Norman, John 19
Northampton 121
Northumberland, Henry Percy, 4th Earl of 170, 177
Norwich 6, 24
Notary, Julian 188
Nyche, Thomas 10, 11, 13

Offord, M. Y. 126-7
Orton-on-the-Hill, Warw. 13 *n*2
Os, Gotfridus de 160, 182
Otranto 105, 115
Overey, William 25-31, 33, 38, 41, 42-4
Ovid 176; *Métamorphoses moralisées* 73, 74, 76 *n*2, 80 *n*3, 101-2, 156*n*; *see also* Caxton: *Ovid, Metamorphoses moralised*
Oxford, John de Vere, 12th Earl of 6*n*, 114*n*, 165; John, 13th Earl of 6*n*, 65, 67 *n*1, 80, 164-7, 169, 170;

Robert 3rd Earl, Robert 5th Earl, Robert 6th Earl, Robert 9th Earl 165 *n*2
Oxford 93, 101, 136, 139-40, 145, 157, 187

Padua 113
Pamping, John 169
Pannartz, Arnold 92 *n*4
Paris 53, 62 *n*2, 78 *n*2, 158, 181
Paston, Anne 169; Sir John 36, 65 *n*2, 112; family 80 *n*2, 87 *n*2, 112, 169
Pénitence d'Adam 74 *n*1 & 3, 76
Pepys, Samuel 53, 102 *n*2
Pepysian Library, Magdalene College Cambridge 102 *n*2, 163
Perceforest 143 *n*1
Pfister, Albrecht 54*n*
Philibert of Savoy 36
Philip the Good, Duke of Burgundy 16, 22, 26, 29-36, 52, 64, 72, 74, 80, 85, 113
Philip the Hardy, Duke of Burgundy 78
Philip III le Hardi, King of France 155
Philip IV le Bel, King of France 155 *n*2
Philippi, Nicolaus 138-9, 187*n*
Phillipps, Sir Thomas 102 *n*2
Pickering, John (d. 1448) 27*n*; John (d. 1498) 18, 27, 28, 34, 37, 43, 44, 50, 67; William (livery 1452) 18, 27*n*; William (executor 1447) 27*n*
Picquigny 69, 119 *n*3
Pierpont Morgan Library, New York 120 *n*3, 136 *n*2, 137 *n*1
Pisan, Christine de *see* Christine de Pisan
Pius II, Pope, 105, 113
Planudes 138
Plasiis, Petrus de 103 *n*1
Plomer, Henry R. 8, 17 *n*2, 93 *n*2, 189
Poggio 138-9, 177
Pontefract Castle 121
Pope, Alexander 64
Poupet, Guillaume de 72
Power, Eugene B. 102 *n*2
Pratt, William 18, 92, 155-6, 187
Premierfait, Laurent de 86 *n*2, 112-13
Preste, Simon 35
Printer of *Abusé en cour* 137 *n*2; of *Albertus Magnus* 56 *n*2; of *Augustinus de Fide* 57, 59 *n*3; of *Dares* 56-7; of *Dictys* 56-7; of *Flores sancti Augustini* 56-7, 60, 75, 76 *n*4; of *Haneron* 75 *n*4

Procida, Johannes de 87
Proctor, Robert 136, 156, 157
Proust, Marcel 30 *n*1, 70
Prout, John 37
Purde, William 116
Pynson, Richard 83, 96 *n*7, 103,
    116 *n*3, 134, 140 *n*1, 173 *n*1, 179,
    188, 190

Rabelais, François 64
Ratcliffe, Sir Richard 129
Raynton, John 99
Reaney, P. H. 2 *n*1
Rede, John 20
Redeknape, Emond 18, 191; Richard
    191; William 18, 34, 37, 191
Regimen sanitatis 180
Reinhart, Marcus 138-9
Rhodes 105, 115
Richard II 130 *n*1, 165 *n*2
Richard III 29-30, 36, 41, 46, 47, 65,
    69, 71, 86 *n*1, 89 *n*4, 108, 114,
    116, 121-4, 129, 135, 136 *n*2,
    138-47, 149-52, 164, 166, 167,
    168 *n*3, 173
Richard, Duke of York, father of
    Edward IV 6*n*, 25, 65, 84 *n*2, 151
Richard, Duke of York, s. of Edward
    IV 90, 115, 121-3, 129*n*, 191
Richard the Lovelorn, jester 37
Rickling Hall, Essex 83 *n*3
Ridgway, Maurice H. 160 *n*3
Ripon Cathedral Library 104 *n*4
Rivers, Anthony Woodville, 2nd Earl
    5, 32*n*, 36, 37, 40-2, 45, 47,
    65 *n*2, 67, 84-91, 94 *n*1, 95-6, 106,
    112-116, 121-3, 131, 142, 145,
    147, 165, 167-9, 187
Rivers, Richard Woodville, 1st Earl
    32*n*, 39, 86, 114
Robert of Shrewsbury 157
Robinson, Lionel 102 *n*2; Philip
    102 *n*2
Rochester, Kent 4, 26 *n*1
Rolewinck, Werner 106 *n*1, 117, 188
Rome 53, 63, 92 *n*4, 101
Romney, Kent 4, 5
Romuleon 72
Rood, Theodoric 136, 139-40, 157,
    183
Roos, Mary 187
Rosenbach Foundation, Philadelphia
    116 *n*1
Rosse, William 83 *n*2
Rotheram, Thomas 71 *n*2, 121 *n*2
Rotterdam 68
Royal Library, Windsor 172
Roye, Guy de 170 *n*2

Ruppel, Berthold 54*n*
Russell, John 35, 37, 40-1, 45, 46, 50,
    67, 94-5, 121, 135; *see also*
    Caxton: *Propositio Johannis
    Russell*
Rutland, Edmund, Earl of 84 *n*2
Ruysschaert, J. 97

St. Albans, Herts., first battle of 30 *n*2;
    second battle of 29-30; Benedic-
    tine Abbey at 113 *n*3, 179, 183;
    Grammar School 120; School-
    master Printer at 93 *n*2, 101, 134,
    139-40, 157, 179, 188
St. Aubin-du-Cormier 167
St. Bride's Institute Library 99*n*
St.-Omer 31 *n*3, 32, 34, 35
Salerno 96 *n*2
Salford, John 71
Salisbury, Sir Thomas Montagu, Earl
    of 13
Sandwich, Kent 4, 5, 32*n*
San Gemignano 154
Sant, John, Abbot 84, 104, 164, 183
Saona, Laurentius Traversagni de 96-
    97, 104, 135
Sawston, Thomas 5 *n*1
Saye, James, Lord 16
Scales, Lady Anne 89; Anthony Wood-
    ville, Lord *see* Rivers; Thomas,
    Lord 89 *n*1
Scarborough 24, 44 *n*1
Schenck, Peter 80 *n*3
Schilling, Johann 56-7, 80 *n*3
Schoeffer, Peter 35, 54*n*, 75, 76
Schoolmaster Printer *see* St. Albans
Scott, E. J. L. 5, 93 *n*2; Sir John 68,
    81
Scrope, Millicent 114*n*; Stephen 87 *n*2,
    112, 114*n*; Sir Stephen 114*n*
Selle, John 17, 18 *n*2
Selley, Cecily 100; David 100, 101
Sforza, Galeazzo, Duke of Milan 36
Shaa, Friar 138
Shackerstone, Leics. 13 *n*2
Shakespeare, William 16*n*, 25 *n*2,
    30 *n*1, 105, 112
Shamwell, Hundred of, Kent 4
Sheffelde, mercer 176
Shelley, John 18, 20
Sheppard, Leslie A. 56 *n*1, 72 *n*2, 73
Shirwood, John 136 *n*2
Shore, Jane, née Elizabeth Lambert
    168-9, 173; William 20, 169, 173
Shrewsbury Abbey 157
Shuckburgh, Sir John 116 *n*2
Shukburghe, Thomas 63*n*
Simnel, Lambert 164, 165 *n*5,167,191

Sixtus IV, Pope 83-4, 96, 102, 104-6, 115, 135-6, 168 *n*2, 169, 178
Skeat, W. W. 2 *n*1
Skelton, John 176-7
Socrates 87-9
Somerset, Edmund Beaufort, 2nd Duke of 30, 116; Henry, 3rd Duke 30; Edmund, 4th Duke 36, 42, 46
Somme le Roi 155, 170; *see also* Caxton: *Royal Book*
Somme rurale 76 *n*2
Sonnyng, William 161
Sotheby and Co. 102 *n*2, 106
Southampton 17, 25, 28, 87, 113
Spenser, Edmund 92 *n*1
Stacy, John 65, 90
Stafford, Sir Humphrey 164
Standon, Herts 160 *n*3
Stanley, Thomas 116, 151; Sir William 116, 149, 151
Stanney, John 144
Staunton, Thomas 12 *n*1, 14 *n*2
Steeple Bumpstead, Essex 6*n*
Steinhöwel, Heinrich 138-9
Steinschaber, Adam 148 *n*
Stevenson, Allan H. 56 *n*2, 59-60, 85, 120, 135
Stockton, John 31, 44, 47
Stoke, battle of 164, 165, 167
Stondo, Bernard van 140 *n*3
Stow, John, *Survey of London* 11, 14, 174 *n*2
Stowell, Robert 190
Strassburg 54, 55*n*, 92
Streete, Randolph 10, 12
Strood, Kent 4, 26 *n*1
Strozzi, Jacques 26
Subiaco 53
Sulpitius, Johannes, Verulanus 93
Surigonus, Stephanus 92-3, 96, 135
Suso, Henricus 182
Sweynheym, Conrad 92 *n*4

Tafur, Pero 22, 23, 52
Tanner, L. E. 82, 100
Taylor, John 144 *n*3
Ten Raem, Gerard 101
Tenterden, Kent 4, 5, 7
Tewkesbury, battle of 46-7
Ther Hoernen, Arnold 54-7, 76 *n*4, 78
Thielemans, M. R. 19
Thomas, Alan G. 134 *n*2
Thomson, Roy, Lord 6
Thorney, Roger 116
Tignonville, Guillaume de 87

Tiptoft, John, Earl of Worcester *see* Worcester; Robert, 3rd Lord 114*n*
Tottel, Richard 179
Toulouse 57
Towton, battle of 30 *n*1, 32
Traversagni, Laurentius, de Saona *see* Saona
Treatise of Love 189
Trevisa, John 106, 117-18
Turnat, Richard 14 *n*1
Tutet, Mark C. 116 *n*2
Twynho, Ankarette 90
Tynemouth, John of 117 *n*3
Tyre, William of 114

Ulm 138
Urban VI, Pope 103
Urswick, Christopher 129
Utrecht 33, 35, 39, 48, 54, 80, 103, 124 *n*4

Valenciennes 31
Valerius Maximus 76 *n*7
Van der Linden, H. 46
Van der Mye, Ghisbert 68-9
Van Praet, J. B. B. 72 *n*2
Vatican Library 97, 104 *n*2
Vaughan, Richard 33 *n*1, 40 *n*2, 85; Sir Thomas 35, 37, 45, 121 *n*3
Vegetius 169
Veldener, Johann 55-62, 73-7, 79, 80, 86, 95, 101, 117 *n*2, 124, 140 *n*1, 152, 160 *n*2, 163, 168, 181, 182, 189 *n*2
Venice 22, 53, 96 *n*2, 113 *n*2, 135-6, 158 *n*2
Vienne, Dauphiny 57, 80 *n*3, 150
Vignay, Jean de 64, 65, 145
Villon, François 91
Vinaver, Eugene 147-8
Vincent of Beauvais 148
Virgil 52, 92, 174, 176; *see also* Caxton: *Eneydos*
Vitas patrum 184, 187
Vliederhoven, Gerard van 80
Voragine, Jacobus de 145; *see also* Caxton: *Golden Legend*

Wakefield, battle of 25, 84 *n*2
Wallingford, William, Abbot 179, 183
Walsh, William, Prior 100
Wanmate, Herman 68
Warbeck, Perkin 65, 105 *n*, 149, 191
Ward, Richard, priest 190
Wareyn, John 29
Warton, Thomas 94

Warwick, Richard Neville, Earl of, the Kingmaker 26, 30-47, 49, 51, 65, 66, 67 *n*1, 86, 106, 113, 114, 122, 144 *n*1 & 2, 153*n*, 165
Wayneflete, William 112, 140
Webb, Christopher A. 125-6
Weiss, Roberto 136 *n*1
Wenlock, John, Lord 34, 37, 45, 47, 49
Wenssler, Michael 158 *n*1
Westfalia, Johannes de 59 *n*3, 61*n*, 124 *n*4, 140 *n*3
Westminster Abbey Library 80
Weston, Sir John 115
Whetehill, Sir Richard 32-4, 49
Willemszone, Pieter 29
Wilson, R. H. 139 *n*1
Winchester 147, 157 *n*2, 167; College Library 147
Windsor 121; *see also* Royal Library
Winnington, Robert 17
Woodstock, Thomas of 129 *n*
Woodville, Anne 93 *n*3, 137; Anthony *see* Rivers; Catherine 129*n*; Sir Edward 167; Elizabeth *see* Elizabeth, Queen, wife of Edward IV; Sir John 39, 86; Lionel 123*n*; Richard *see* Rivers
Worcester, John Tiptoft, Earl of 35 *n*2, 41, 112-14, 165
Worcester, William 112-13
Worde, Elizabeth de 98; Julian de 98*n*; Wynkyn de 54, 61, 70, 82, 83, 84, 96 *n*7, 98, 103 *n*2, 116, 118, 134, 140 *n*1, 150, 153*n*, 154, 160, 166 *n*1, 173 *n*1, 178, 179, 182-90
Würzburg 56
Wyche, Hugh 31
Wydeville *see* Woodville
Wynkyn de Worde *see* Worde

Yale University Library 123
York 24, 44 *n*1
Ypres 17, 38

Zainer, Johann 138
Zel, Ulrich 54, 57 *n*1